MATHEMATICAL JOURNEYS

Peter D. Schumer
Middlebury College

A JOHN WILEY & SONS, INC., PUBLICATION

Published by John Wiley & Sons, Inc., Hoboken, New Jersey.
Published simultaneously in Canada.

For general information on our other products and services please contact our Customer Care Department within the U.S. at 877-762-2974, outside the U.S. at 317-572-3993 or fax 317-572-4002.

Wiley also publishes its books in a variety of electronic formats. Some content that appears in print, however, may not be available in electronic format.

Library of Congress Cataloging-in-Publication Data:

Schumer, Peter D., 1954–
Mathematical journeys / Peter D. Schumer.
p. cm.
"A Wiley-Interscience publication."
Includes bibliographical references and index.
ISBN 0-471-22066-3 (pbk. : acid-free paper)
1. Mathematics—Popular works. I. Title.

QA93.S38 2004
510–dc22

2003062040

Printed in the United States of America.

10 9 8 7 6 5 4 3 2

I dedicate this book to Amy and Andrew who are well on their way with their own wonderful life's journeys.

Contents

Preface

Imagination is as good as many voyages—and how much cheaper.

—George William Curtis (1824–1892)

For as long as I have taught college mathematics, I have been convinced that despite personal protestations to the contrary, a great many people find real mathematics to be quite interesting and very appealing. In addition, with some patience and a willing spirit most are fully capable of understanding sophisticated mathematical discourse. If the basic ingredients are few and simple and the problems straightforward and clear, then natural human curiosity provides all the motivation necessary to work through all the steps of a rigorous argument. In mathematics there is a keen sense of accomplishment and finality derived from a result cleanly proved that is nearly unique among intellectual enterprises. The thrill of discovery and deep understanding that motivates youngsters to find out "what comes after addition and subtraction" also drives professional mathematicians in their state-of-the-art research.

In this book I hope to present genuine mathematics to anyone who wants a real taste of what mathematicians consider fascinating and beautiful. The book is generally laid out buffet style, so feel free to jump in anywhere and sample whatever tempts you. If you read most of the book you'll have had a fairly nutritious meal of mathematics. I assume that you have had the usual high school mathematics, but I will not assume that you were always thrilled by it. Knowing basic rules of algebra (like add the same thing to both sides of an equation) and standard geometric definitions (e.g., knowing that a quadrilateral has one more side than a triangle) are nearly sufficient prerequisites. The rest will be introduced to you as needed in the text (even if you already know some of it).

My teaching philosophy has always been to begin with basic and elementary notions in order to make sure everyone is on board. Then I move along from simple notions gradually toward more complex and hopefully more interesting ones. Every class should include a worthwhile discovery or a memorable result. I apologize at the beginning to those students who feel they've seen some of the material before, but I promise them that the rewards are great for those who can understand mathematics solidly and well. After all, don't professional baseball players limber up and practice the basics of throwing and hitting every spring to warm up for the real season? Similarly, one must be completely comfortable with the mathematical foundation upon which grander results are built. So each chapter moves (smoothly I hope) from basic background toward some really neat and worthwhile mathematics.

Over the years, I have given about two dozen general mathematical lectures at Middlebury College as part of the seminar series in my department. I have strived to make those lectures both accessible to new college students and also of some interest to the experienced mathematicians in the audience. In writing this book, I have borrowed from many of these lectures and reworked them to give a more even treatment here. Very little of what I present is original mathematics (i.e., created by me and presented for the first time here.) Instead I have collected material from a wide variety of disparate sources from which I have drawn connections to make a cohesive whole. The history of ideas and the details of the lives of the people behind the theorems also interest me. I include a fair amount of historical background when I think it enhances the exposition, but I have avoided long historical discourses or philosophical asides when they detract from the presentation. What I have done in some sense, is to organize a colorful tour for you through the fascinating and beautiful land of mathematics. There are chapters on primes, on various mathematical games, on infinite series, on calculating pi, unusual geometric problems, variations on the partition function, and much more. Whether you are a high school or college student who wants to know a bit about the sort of mathematics that's just not covered in a standard calculus course, an adult who would like to learn more about the real essence of mathematics, or a college mathematics instructor who wants to add a bit of spice to your courses, I hope there is much here to please you. There are some great panoramic views ahead. Have an enjoyable and stimulating journey!

One final comment: Mathematics is a vibrant, living organism that continues to move about and grow. This is especially apparent in the area of computational number theory where new results proliferate at such a rate that any publication such as this will be somewhat out of date the moment it is printed. Such an update is in order for Chapter 5. In particular, I am pleased to announce the discovery of a new "largest known" prime on November 17, 2003. The prime in question is $2^{20996011} - 1$, the 40th Mersenne prime comprising 6320430 digits. The prime was discovered through the collective efforts of GIMPS (the Great Internet Mersenne Prime Search). The key participants in this discovery were George Woltman, Scott Kurowski, and Michael Shafer.

Acknowledgments

I would like to express my deep appreciation to Susanne Steitz, Assistant Editor for Mathematics and Statistics, for all her help in ushering my book from an incomplete manuscript to its final stages. I would like to acknowledge Heather Haselkorn, Editorial Program Coordinator, for her help in prodding me in the earliest stages of this work. Special thanks goes to Associate Managing Editor Danielle Lacourciere who worked closely with me during the final stages of the book's production and who helped make the transition from manuscript to published book a smooth and enjoyable one.

For help with obtaining portraits of mathematicians, thank you to G. Dale Miller, Development Associate and Permissions Coordinator at the Smithsonian Institution Libraries. The portraits of Archimedes (page 103), Johann Bernoulli (page 52), Leonhard Euler (page 44), Pierre de Fermat (page 62), Benjamin Franklin (page 74), Carl Friedrich Gauss (page 110), Gottfried Wilhelm Leibniz (page 154), Blaise Pascal (page 165), George Bernhard Riemann (page 142), James Joseph Sylvester (page 158), and Francois Viète (page 104) are provided courtesy of the Dibner Collection of the Smithsonian Institution. The portrait of Srinivasa Ramanujan (page 111) is a reproduction of a 1962 Indian stamp. The photograph of Paul Erdös (page 139) was taken from George Csicsery's documentary film *N is a Number: A Portrait of Paul Erdös*. In addition, the picture of the Green Chicken plaque (page 14) was taken by Frank Swenton shortly before the bird flew south to Williamstown for the winter.

My heartfelt appreciation goes to all my friends, colleagues, and especially the students who have actively listened to me speak on the various topics included here over the past twenty years. Their questions, comments, and generous support help make this all the more fun. Finally, I wish to express my profound gratitude to Lucy, Amy, and Andrew who will always be the ones who really count.

1 Let's Get Cooking: A Variety of Mathematical Ingredients

We begin our mathematical journey by introducing (or reminding you about) some basic objects of mathematics. These include prime numbers, triangular numbers, and geometric squares. We will also present a couple of neat proofs and discuss a very useful tool for carefully establishing results dealing with the natural numbers, namely mathematical induction. This chapter lays the foundation for many of the latter chapters and hence might be a bit more elementary. On the other hand, don't worry about being bored. Mathematics asks (and often answers) interesting fundamental questions. We'll get to some really compelling things right away!

Let's avoid a philosophical discourse on the construction of (or the a priori existence of) the natural numbers. I assume that the natural numbers exist (at least in our humanly defined mathematical world). They are the counting numbers $1, 2, 3, \ldots$ and they continue forever in the sense that if N is a natural number, then so is $N+1$. We denote the set of all natural numbers by $\mathbb{N}$. The set of natural numbers can be subdivided. The number 1 is special in that it is the unique multiplicative identity. This is just a fancy way to say that anything times 1 is itself.

Next come the *primes*. You no doubt recall that these are natural numbers that are only divisible by 1 and themselves. For example, 2 is prime. So are 17 and 41 since they cannot be factored further. So is 10,123,457,689, the smallest prime containing all ten decimal digits. But $6 = 2 \cdot 3$ and $91 = 7 \cdot 13$ are not prime. Numbers like 6 and 91 belong to the third category, namely the *composites*. The primes are the multiplicative building blocks for all the natural numbers. The composite numbers are those that require at least two building blocks.

There are endless questions to be asked about the natural numbers, especially the primes. The Fundamental Theorem of Arithmetic states that all natural numbers have a unique factorization as a product of primes (discounting order of the factors). Since $2{,}002 = 2 \cdot 7 \cdot 11 \cdot 13$, no other product of primes could equal 2,002. But how many primes are there? Are there any consisting of more than a million digits? How many are of the form N^2+1, that is, exceed a square number by 1? How spread apart are they? Some of these questions were answered by

Mathematical Journeys, by Peter D. Schumer
ISBN 0-471-22066-3

the ancient Greek geometers more than 2000 years ago. Others still have not been answered fully. Let's get right to it and answer the first question, which is perhaps the most fundamental of all.

Theorem 1.1: *There are infinitely many primes.*

How does one go about demonstrating the veracity of the statement above? Certainly any list of primes, no matter how long, will not suffice to show that there are *infinitely* many primes. Perhaps showing that the primes satisfy some pattern would help. For example, if for each prime it was true that its double plus 1 was prime, then we'd have a process for producing as many primes as we wished. Let's give it a try by carrying out this process on the first prime: the sequence begins $2, 5, 11, 23, 47, \ldots$ —look promising? So far so good, but the next number in our sequence is 95, a composite number. This isn't going to be that easy. But perhaps a similar idea might work.

A theorem is a mathematical statement whose veracity has been rigorously demonstrated to the satisfaction of mathematicians. At least, that's my definition of a theorem. If it sounds a bit tenuous, it no doubt is. On very rare occasions, a result is erroneously considered to be a theorem for many years before a logical weakness or flaw is discovered in its purported proof. But such instances are extremely unusual. Although mathematics is a human endeavor, mathematicians treat their work very stringently and the standard for proof is high. So from here on you are a mathematician who is about to read the following proof presented by Euclid in *The Elements* (Book IX, Proposition 20), written about 300 B.C.E. (Before the Common Era). See if it convinces you.

Proof of Theorem 1.1: *There are some primes, for example, 2 is prime. So now consider a nonempty set of primes, $S = \{p_1, p_2, \ldots, p_n\}$. The number $N = p_1 \cdot p_2 \cdot \ldots \cdot p_n + 1$ is not divisible by p_1 because division by p_1 leaves a remainder of 1. Similarly, N is not divisible by any of the primes in the set S. So N is either another prime not among those listed in S or N is composite, but made up of a product of primes none of which are in the set S. So the set S is not a complete list of all the primes. Since S was an arbitrary set of primes, no finite list of primes can be complete. Therefore, the number of primes is infinite.* □

The result is stunning and the proof is elegant. We have answered a simple but deep question about the prime numbers and hence about the natural numbers themselves. The natural numbers have an infinite number of multiplicative building blocks. We won't run dry! And notice that we've also answered our second question. Are there any primes consisting of more than a million digits? Yes, there are. In fact, there are infinitely many of them (since only finitely many primes could possibly have a million or fewer digits.) Naming such a prime may be quite a different matter, but later we will tell some of that story as well.

The primes begin 2, 3, 5, 7, 11, 13, 17, 19, 23, 29, 31, 37, 41, 43, 47. There are various sized gaps between successive primes. For example, the gaps from our list above are 1, 2, 2, 4, 2, 4, 2, 4, 6, 2, 6, 4, 2, 4. Since two is the only even prime number, all gaps after the initial gap of 1 are even numbers. How big can these numbers get? Does 2 appear infinitely often? Do all even numbers appear at some point? The next theorem answers our first question.

Theorem 1.2: *There are arbitrarily large gaps between successive primes.*

What we require are arbitrarily long strings of consecutive composite numbers. Below we construct such a string. Before we begin, recall the definition of *factorial*. If n is a natural number, $n!$ (read "*n* factorial") is the product of all the natural numbers up to and including n. For example, $5! = 1 \cdot 2 \cdot 3 \cdot 4 \cdot 5 = 120$.

Proof of Theorem 1.2: *Given a natural number $N \geq 2$, consider the sequence of N consecutive numbers $(N+1)!+2, (N+1)!+3, \ldots, (N+1)!+N+1$. Note that 2 divides $(N+1)!$ since 2 is one of the factors in the product that defines $(N+1)!$. So 2 divides $(N+1)!+2$ and hence $(N+1)!+2$ is composite. Similarly, 3 divides $(N+1)!+3$ and so $(N+1)!+3$ is composite as well. Analogously, all the N consecutive numbers from $(N+1)!+2$ to $(N+1)!+N+1$ are composite. Since the number N is arbitrary, there are strings of consecutive composite numbers of any given length. Hence there are arbitrarily large gaps between successive primes.* □

Ponder for a moment what Theorems 1.1 and Theorems 1.2 say. There is an unlimited supply of primes, but gaps between them can be as large as you like. Also note what the theorems do not claim. Theorem 1.2 does not say that there is an infinite string of consecutive composites. For if there were, then there would have to be a last prime, contradicting Theorem 1.1. Nor does Theorem 1.2 imply that our construction gives the first instance of a particularly sized gap. However, both theorems do tell us something significant about prime numbers.

On the one hand, there are lots and lots of primes. On the other hand, they're not so dense in the natural numbers that there aren't long patches without them. Compare this with the odd numbers. There are infinitely many odd numbers, but there's never a gap larger than two between them. How about the set of squares—1, 4, 9, 16, etc.? Certainly there are infinitely many of them. And the gaps are successive odd numbers that get arbitrarily large. So in some sense the primes are a bit more like the squares perhaps. Deeper theorems do distinguish between the density of squares and the density of primes among the natural numbers, but for now at least we have another concrete model where results analogous to Theorems 1.1 and 1.2 apply. (For those of you who are curious, there are more primes than squares in some strict analytical sense.)

What about our other questions? It has not been proven that every possible even gap appears somewhere along the endless list of primes, but number theorists tend to believe that it is so. In fact, A. de Polignac conjectured that every even number appears infinitely often as a gap between consecutive primes (1849). And although there are lots of examples of the smallest gap of two, even a proof that there are infinitely many such *twin primes* remains elusive. However, many gigantic examples have been discovered. Currently the record is the twin pair $33218925 \cdot 2^{169,690} \pm 1$, found by Daniel Papp (2002). Each number is a 51,090-digit prime.

Now let's consider the *triangular numbers*, for each n defined as the sum of the first n natural numbers. Denote them by t_n. For example, $t_4 = 1+2+3+4 = 10$. Geometrically, we can think of t_n as the number of bowling pins with n rows laid out in the usual triangular array. Imagine some giant extraterrestrial species bowling with t_{10} bowling pins. The 46–55 split is a real killer! A most natural question is, "What is t_n?" Is there a nice formula which gives us t_n for all n? Of course there is. Why would I bring it up? The formula is simple and provides us with the classic example in which to introduce the technique of *proof by induction*.

Theorem 1.3: *For all* $n \geq 1$, $t_n = n(n+1)/2$.

The method of mathematical induction is a simple one. It consists of verifying the assertion for the initial value ($n = 1$ in this case) and then showing that if the theorem is true for the case n, then it must be true for case $n + 1$. In this way, it's much like an infinite set of dominoes all lined up in a row. If we knock over the first one and carefully set up all the rest so that each one knocks over the succeeding domino, then eventually any particular domino will fall. This method of proof was widely used by the brilliant French philosopher and mathematician Blaise Pascal (1623–1662) and so this method of proof is often attributed to him. However, recently mathematical historians have discovered that the same technique was utilized by Levi ben Gerson (1288–1344), the Jewish medieval Biblical scholar, astronomer, philosopher, and mathematician. No doubt the clever idea of mathematical induction has been independently discovered by many scholars. Its elegance and simplicity belie its power and broad applicability.

Proof of Theorem 1.3: *For* $n = 1$, $t_1 = 1 = 1(2)/2$. *So Theorem 1 is correct in this case. If the formula is correct for* t_n, *then let's show that it is correct for* t_{n+1} *as well.*

$$\begin{aligned} 1 + 2 + \ldots + n + (n+1) &= (1 + 2 + \ldots + n) + (n+1) \\ &= n(n+1)/2 + (n+1) \text{ by the inductive hypothesis} \\ &= [n(n+1) + 2(n+1)]/2 \\ &= (n+1)(n+2)/2, \end{aligned}$$

which is the formula for t_{n+1}. *By mathematical induction, the formula holds for all* $n \geq 1$. □

Make sure to double-check the little bit of algebra in the proof above and be certain that you are comfortable with the logic behind mathematical induction. If so, you are well on your way toward thinking like a real mathematician. Now if you need to know the sum of the first 1,000 natural numbers, the solution is easy: $t_{1,000} = 1{,}000(1{,}001)/2 = 500{,}500$. Watch out for that bowling ball!

Let's try our hand on a somewhat more intricate example. Define the *Fibonacci numbers* as the numbers in the sequence 1, 1, 2, 3, 5, 8, 13, 21, 34, etc. Each entry in the sequence from the third entry on is the sum of the two preceding Fibonacci numbers. Hence $F_1 = 1$, $F_2 = 1$, and $F_n = F_{n-1} + F_{n-2}$ for $n \geq 3$. The Fibonacci sequence is a good example of a sequence defined *recursively*. These numbers are named after the Italian mathematician Leonardo of Pisa (ca. 1175–1250), who promoted the use of the Hindu-Arabic numeral system. (He was also known as Fibonacci, son of Bonaccio.) His book, *Liber Abaci* (1202), contained the following problem: "How many pairs of rabbits can be produced from a single pair in a year if every month each pair begets a new pair which from the second month on becomes productive?" The sequence thus produces the number of pairs we've listed above.

There seem to be an endless number of interesting arithmetic formulae based on this simple sequence. For example, let's pick any Fibonacci number, square it, subtract from that the product of its two neighboring Fibonacci numbers (above and below), and see what results. If we choose $F_7 = 13$, then we get $(F_7)^2 - (F_6 \cdot F_8) = 13^2 - 8 \cdot 21 = 1$. If we choose $F_8 = 21$, then a similar computation gives us $21^2 - 13 \cdot 34 = -1$. Experiment with some other starting values. Really—try some other examples. Once you've done a few calculations, then read on. (I'm trying to help get you in the mindset for mathematical discovery. This is the essence of mathematics.)

No doubt you have noticed that the answer is always 1 or -1. In fact, if we pick an odd-indexed Fibonacci number we get $+1$ and if we choose an even-indexed Fibonacci number we obtain -1. This leads us to the conjecture that $F_n^2 - F_{n-1} \cdot F_{n+1} = (-1)^{n-1}$. In fact, this is a theorem that we now prove in two different ways, each making use of mathematical induction.

Theorem 1.4 (J.D. Cassini, 1680): *For all $n \geq 2$, $F_n^2 - F_{n-1} \cdot F_{n+1} = (-1)^{n-1}$.*

First Proof of Theorem 1.4 (Direct): *Our initial value is $n = 2$: $F_2^2 - F_1 \cdot F_3 = 1^2 - 1 \cdot 2 = (-1)^1$. So the proposition is true in this case. Now let's assume that it holds for some unspecified value of n, that is, $F_n^2 - F_{n-1} \cdot F_{n+1} = (-1)^{n-1}$. We will show that the proposition necessarily follows in the next case, namely for $n + 1$.*

$$F_{n+1}^2 = F_{n+1}(F_n + F_{n-1}) \text{ by the definition of } F_{n+1}$$

$$= F_{n+1}F_n + F_{n-1}F_{n+1} + (F_n^2 - F_n^2)$$

(mathematicians love to add zero to equations since nothing gets disturbed)

$$= F_n^2 + F_{n+1}F_n - (F_n^2 - F_{n-1}F_{n+1})$$

$$= F_n(F_n + F_{n+1}) - (F_n^2 - F_{n-1}F_{n+1})$$

$$= F_n F_{n+2} - (F_n^2 - F_{n-1} F_{n+1}) \text{ by the definition of } F_{n+2}$$

$$= F_n F_{n+2} - (-1)^{n-1} \text{ by the inductive assumption.}$$

Hence $F_{n+1}^2 - F_n F_{n+2} = (-1)^n$ *and the result follows.* □

Theorem 1.4 also affords us the opportunity to introduce some basic matrix arithmetic. Simply put, a *matrix* is a rectangular array of numbers. Mathematicians have defined various arithmetic operations on matrices such as addition, subtraction, multiplication, and so on that share many of the standard properties of the ordinary arithmetic operations on real numbers. Here we need only be concerned with 2×2 square matrices consisting of four entries in two rows and two columns. Matrix addition and subtraction is defined component-wise and appears perfectly natural. For example, if

$$A = \begin{bmatrix} 1 & 3 \\ 5 & 7 \end{bmatrix} \text{ and } B = \begin{bmatrix} 2 & 4 \\ 6 & 8 \end{bmatrix},$$

then their sum

$$A + B = \begin{bmatrix} 1+2 & 3+4 \\ 5+6 & 7+8 \end{bmatrix} = \begin{bmatrix} 3 & 7 \\ 11 & 15 \end{bmatrix}.$$

Multiplication is not defined component-wise, but rather in a bit more intricate way. If

$$M = \begin{bmatrix} a_1 & b_1 \\ c_1 & d_1 \end{bmatrix} \text{ and } N = \begin{bmatrix} a_2 & b_2 \\ c_2 & d_2 \end{bmatrix}$$

are any two such matrices, then we define their product to be the 2×2 matrix

$$MN = \begin{bmatrix} a_1 a_2 + b_1 c_2 & a_1 b_2 + b_1 d_2 \\ c_1 a_2 + d_1 c_2 & c_1 b_2 + d_1 d_2 \end{bmatrix}.$$

For example, with A and B defined as before, their product

$$AB = \begin{bmatrix} 1 \cdot 2 + 3 \cdot 6 & 1 \cdot 4 + 3 \cdot 8 \\ 5 \cdot 2 + 7 \cdot 6 & 5 \cdot 4 + 7 \cdot 8 \end{bmatrix} = \begin{bmatrix} 20 & 28 \\ 52 & 76 \end{bmatrix}.$$

In addition, we define positive powers of a matrix by $M^2 = MM$ and, more generally, $M^n = MM^{n-1}$ for $n \geq 2$.

Next we define the determinant of M. If

$$M = \begin{bmatrix} a & b \\ c & d \end{bmatrix},$$

then the *determinant* of M, denoted by det M, is the number $ad-bc$. For example, with A as previously defined, det $A = 1 \cdot 7 - 3 \cdot 5 = -8$. Determinants play an important role in linear algebra where matrices are used extensively to study linear transformations. A key result is that the determinant function is multiplicative. That is, if M and N are two 2×2 matrices, then $\det(MN) = \det M \cdot \det N$, a fact you may wish to verify directly. The somewhat complicated nature of the product of two matrices is more than compensated by the simple nature of the product of their determinants.

Second Proof of Theorem 1.4 (Determinants): *We begin by extending our definition of Fibonacci numbers to include* $F_0 = 0$. *Notice that* $F_0 + F_1 = 0 + 1 = 1 = F_2$, *which is consistent with our Fibonacci recurrence definition. Let*

$$M = \begin{bmatrix} 1 & 1 \\ 1 & 0 \end{bmatrix} = \begin{bmatrix} F_2 & F_1 \\ F_1 & F_0 \end{bmatrix}.$$

By induction,

$$M^n = \begin{bmatrix} F_{n+1} & F_n \\ F_n & F_{n-1} \end{bmatrix}.$$

To see this, note that the assertion is consistent with M^1, *our base case. Assuming that*

$$M^n = \begin{bmatrix} F_{n+1} & F_n \\ F_n & F_{n-1} \end{bmatrix},$$

we compute M^{n+1}. *We have that*

$$M^{n+1} = MM^n = \begin{bmatrix} F_{n+1} + F_n & F_n + F_{n-1} \\ F_{n+1} & F_n \end{bmatrix} = \begin{bmatrix} F_{n+2} & F_{n+1} \\ F_{n+1} & F_n \end{bmatrix}$$

as desired. But the determinant is multiplicative. Hence $\det(M^n) = (\det M)^n$ *where* $\det M = -1$. *It follows that* $F_{n+2}F_n + F_{n+1}^2 = (-1)^n$. *Equivalently,* $F_n^2 - F_{n-1} \cdot F_{n+1} = (-1)^{n-1}$. □

Theorem 1.4 shows that the square of a Fibonacci number and a product of neighboring Fibonacci numbers are nearly equal, in fact just missing by one. This fact is the basis of a very compelling geometric paradox attributed to the Oxford don Charles L. Dodgson (1832–1898), better known as Lewis Carroll of *Alice in Wonderland* fame. See the diagram in Figure 1.1. The four shapes comprising the two figures match identically and yet it appears that their total areas differ by one unit. The figure on the left is a square with area 64, while the figure on the right appears to be a rectangle with area 65. The solution to this apparent paradox is that the four pieces on the right don't quite fit together neatly. For example, the slope of the two triangles is $3/8 = F_4/F_6$ while that of the appropriate side of the light and dark trapezoids is $2/5 = F_3/F_5$ (nearly the same, but not exactly). The visual paradox makes use of the fact that the ratio F_{n-1}/F_{n+1} converges fairly quickly so that it's difficult to see that there are actually two broken diagonal lines rather than a single diagonal.

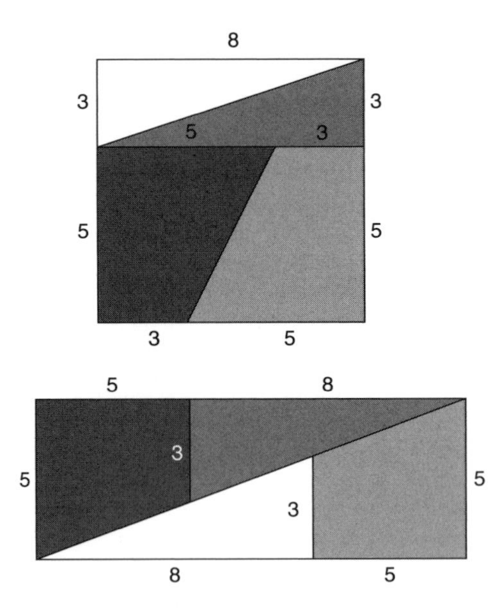

Figure 1.1 Apparent rearrangement of four figures results in different areas.

Another interesting application of Fibonacci numbers is the representation of positive integers as the sum of Fibonacci numbers F_2, F_3, F_4, etc. In this instance we do not need $F_1 = 1$ since $F_2 = 1$ as well. In fact, even without repeating a given Fibonacci number, there may be several ways to write a given integer as such a sum. For example, $17 = 13 + 3 + 1 = 8 + 5 + 3 + 1$. We can denote these *Fibonacci representations* by $17 = (100101)_F$ or $17 = (11101)_F$ where we read from right to left beginning with F_2 and adding the appropriate Fibonacci number every time a 1 appears. A nice result of Edouard Zeckendorf (1972) states that every natural number has a unique Fibonacci representation with no consecutive 1s. So every positive integer can be expressed as a sum of Fibonacci numbers, no two being consecutive. Appropriately, such a Fibonacci representation is called a *Zeckendorf representation.*

We now turn our attention to a geometric example. Since the principle of mathematical induction is much like an endless array of dominoes, let's see what happens when we apply induction to sets of real dominoes. Actually, we will consider L-shaped dominoes that cover three squares of a chessboard. In addition, we will be interested not just in the usual 8×8 chessboard, but also in chessboards of dimensions $2^n \times 2^n$ for any $n \geq 1$. Remove any single square of such a chessboard. I claim that an appropriate number of such L-shaped dominoes can neatly cover the rest of the board. An actual construction might take some time, but induction easily works to show this claim is valid.

If $n = 1$, it's simple to verify that one L-shaped domino can cover the other three squares once one has been removed. See Figure 1.2. Now assume that the

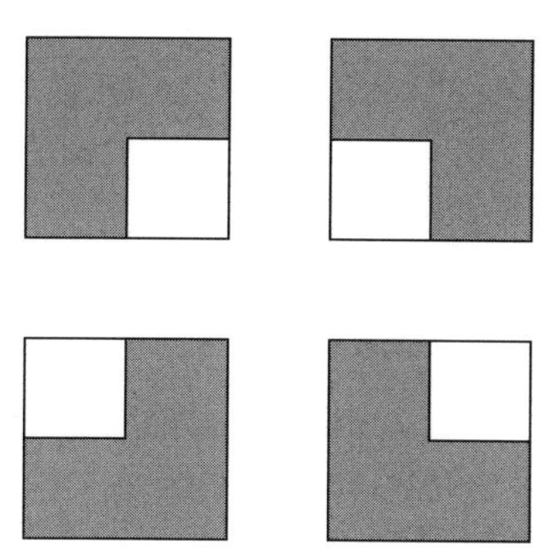

Figure 1.2 Solution for a 2×2 chessboard.

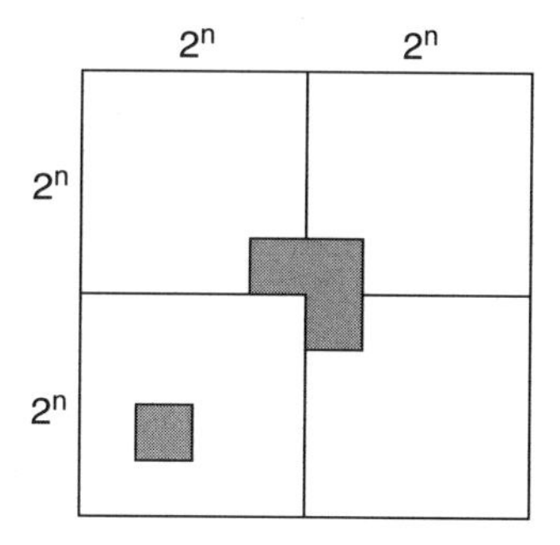

Figure 1.3 Solution for $2^{n+1} \times 2^{n+1}$ chessboard.

result is true for a $2^n \times 2^n$ chessboard with any square taken out. Next consider a $2^{n+1} \times 2^{n+1}$ chessboard. Mentally slice the chessboard in half vertically and half horizontally so that the board is actually made up of four quarter boards, each of dimension $2^n \times 2^n$. If a square is removed from the original large board, it must lie in exactly one of the four quarter boards. By our inductive hypothesis, the rest of that quarter board can be covered with L-shaped dominoes. Now consider the corner where the other three quarter boards meet. Remove the corner square from each of the three quarter boards. Each quarter board with one square removed can be covered with L-shaped dominoes. What remains is one last gap of three squares forming an L-shape. Fill that piece with an L-shaped domino and we are done (Fig. 1.3)!

In the mathematical literature such L-shaped dominoes are often called *right trominoes*. From there it's a small step to objects like tetraminoes and pentominoes and a whole host of new and interesting questions. Such is the nature of mathematical generalization!

Finally, please try your hand at the following cute problem. Show that a given geometric square can be decomposed into n squares, not necessarily all the same

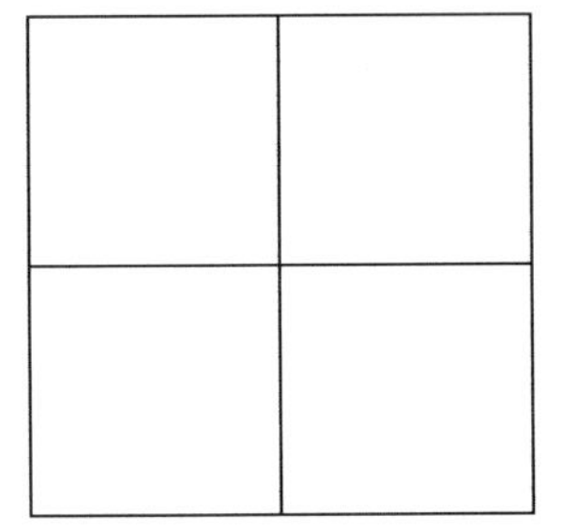

Figure 1.4 Four squares.

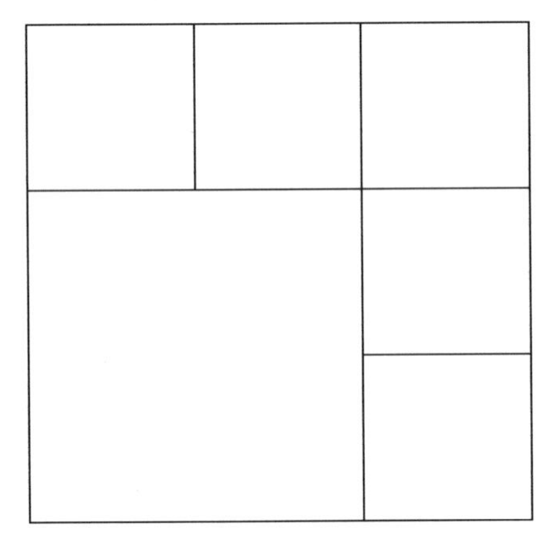

Figure 1.5 Six squares.

size, for $n = 4$ and for all $n \geq 6$. It's usually a good idea to check out some initial cases by hand. Then try to understand *why* the assertion must be true.

The case $n = 4$ is almost trivial. (I don't like to claim that anything actually *is* trivial. I've been puzzled by too many supposedly obvious things over the years to want to pull the same thing on you.) Here's a square broken into four smaller squares, each with sides half the length of the original square (Fig. 1.4). For $n = 6$, the following pattern fits the bill with one square having twice the length and width of the other five squares (Fig. 1.5). And for $n = 8$, a similar idea allows us to break up one square into one large one and seven identical smaller ones, each having one-third the length of the remaining square (Fig. 1.6).

The first picture shows that any square can be broken up into four smaller squares, thereby increasing the number of squares by three. Hence the first square depicted can be decomposed into $4, 7, 10, 13, 16, 19, \ldots$ squares. Notice that these are all the numbers greater than or equal to four having remainder 1 when divided by 3. Similarly, the second square can be decomposed into $6, 9, 12, 15, 18, \ldots$ squares. These are all the numbers greater than or equal to six that are divisible by 3. And the third square can be decomposed into

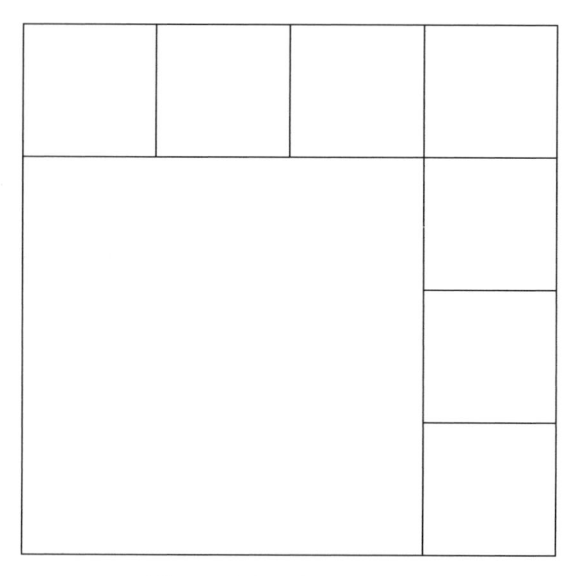

Figure 1.6 Eight squares.

$8, 11, 14, 17, 20, \ldots$ squares. These are all the numbers greater than or equal to 8 that have a remainder of 2 when divided by 3. But all natural numbers are either divisible by 3 or have a remainder of 1 or 2 upon division by 3. Hence a given square can be decomposed into n squares for $n = 4$ and for all $n \geq 6$.

Congratulations, you've completed the first chapter and have digested some serious mathematics. I hope you share my delight in making discoveries about primes, Fibonacci numbers, triangular numbers, and geometric squares. Have you discovered any other interesting properties? Can you prove it?

WORTH CONSIDERING

1. Notice that $2 + 1 = 3$, $2 \cdot 3 + 1 = 7$, and $2 \cdot 3 \cdot 5 + 1 = 31$ are all primes. Investigate how far this phenomenon continues by considering the product of the first n primes plus one.
2. Let p be prime and consider the number $2^p - 1$. Investigate the primality of $2^p - 1$ for various small values of p. (Such numbers are called Mersenne numbers, named after Marin Mersenne (1588–1648), who made some long-ranging conjectures about them.)
3. Fill in the details of the following alternative proof (due to T.L. Stieltjes, 1890) that there are infinitely many primes: Let A and B be two distinct finite nonempty sets of primes. Let P_A and P_B be the product of the elements of A and B, respectively. Consider the factorization of $P_A + P_B$. (For fun and insight, carry out this process on some of your own examples.)

4. Explicitly work through the construction of six consecutive composites in the proof of Theorem 1.2. Find a smaller example of six consecutive composites.

5. **(a)** Let t_n be the nth triangular number. Show that $t_{n-1} + t_n = n^2$.

(b) Show that $t_{n-1}^2 + t_n^2 = t_{n^2}$.

6. Define the nth pentagonal number by $p_1 = 1$ and $p_n = 1+4+\ldots+(3n-2)$ for $n \geq 2$. Why is this a good name for p_n? Use induction to prove that $p_n = n(3n-1)/2$.

7. Verify the following Fibonacci identities:

(a) $F_1 + F_2 + \ldots + F_n = F_{n+2} - 1$.

(b) $F_2 + F_4 + \ldots + F_{2n} = F_{2n+1} - 1$.

(c) $F_1 + F_3 + \ldots + F_{2n-1} = F_{2n}$.

8. Establish that $F_1^2 + F_2^2 + \ldots + F_n^2 = F_n F_{n+1}$.

9. **(a)** Show that the number of ways to fill a bowl with n stones by dropping in either one or two stones at a time is F_{n+1}.

(b) What is the number of ways to cover an $n \times 1$-sized board with an assortment of 1×1 squares and 2×1 dominoes?

10. Verify that if A and B are two 2×2 matrices, then $\det(AB) = \det A \cdot \det B$.

11. Find Zeckendorf representations for the following numbers: 19, 32, 232.

12. Can L-shaped dominoes cover a $3^n \times 3^n$ chessboard with one missing square? What about a $3^n \times 3^n$ chessboard with no missing square? What about a $4^n \times 4^n$ chessboard with one missing square?

13. How many ways can a square be decomposed into ten smaller squares? (There are at least two possible answers depending on whether you count symmetric rotations of a given square as being distinct.)

14. Investigate how many ways a $2 \times n$ chessboard can be covered with n 2×1 dominoes.

15. Primes p for which $2p + 1$ are also prime are called *Germain primes* in honor of Sophie Germain (1776–1831) who demonstrated their intimate connection with Fermat's Last Theorem. Determine how many of the first 25 primes are Germain primes. (No one knows if there are infinitely many such primes.)

16. (L. Euler) Does the following polynomial produce just primes: $f(n) = n^2 - n + 41$?

2 The Green Chicken Contest

Every December, several thousand undergraduate college and university students from Canada and the United States compete in the William Lowell Putnam Mathematical Competition, better known as the Putnam Exam. The contest consists of two arduous sessions, a three-hour morning and a three-hour afternoon period. Students work on six challenging mathematics problems in each session. Tests are graded out of a total of 120 points (10 points per problem). Very little partial credit is granted. The median score is usually zero! Students work individually, but many are members of predetermined institutional teams. The pressure to succeed can be high, but the reward of being on one of the top five winning teams or being one of the five highest ranking individuals (a Putnam fellow) means a lifetime of glory (at least within mathematical circles).

Since 1978, Middlebury College and Williams College have also competed in a less formal pre-Putnam contest. The originators of the contest were two friends and math colleagues, Bob Martin (of Middlebury College) and Peter Andrews (then of Williams College). Each team can field as many students as they wish and no official team is predetermined. In the end, the top four scores from each school are added together and the winning school is, of course, the one with the higher total. The contest consists of six mathematical questions and partial credit for good ideas is given generously. Early on it was determined that the winning trophy would reside with the team that last won the contest. In fact, the winning trophy is a plaque attached to a wooden box surmounted by a rather ugly ceramic green chicken cookie container. There was some lighthearted discussion about whether it wouldn't be more appropriate for the losing team to be stuck with the trophy for a year. Be that as it may, the event is always a lot of fun and a real celebration of mathematical learning and comradeship. The contest has become known affectionately as the Green Chicken Contest.

Here are six problems (and solutions) from previous Green Chicken Contests. Play around and try to solve them yourself before reading the answers. Some of the problems are original to the Green Chicken Contest, while others are old chestnuts passed on from one generation to the next. For example, I have seen the very first problem pop up here and there in different guises. Recently I was able to track it down to the 1901 Hungarian Eötvös Competition, named after the

Mathematical Journeys, by Peter D. Schumer
ISBN 0-471-22066-3

Middlebury and Williams Colleges' Green Chicken trophy.

founder and first president of the Mathematical and Physical Society of Hungary. Maybe the problem is even older.

Problem #1 (1978 Green Chicken Contest): *Prove that, for any positive integer n, $1^n + 2^n + 3^n + 4^n$ is divisible by 5 if and only if n is not divisible by 4.*

Problem #1 was the first problem on the first Green Chicken Contest and serves as a good introduction to the mathematics required to handle such problems. Always begin with some examples to get a feel for the problem and to see if it seems plausible. Remember, no set of examples, no matter how numerous, suffices to establish that an assertion is valid for all natural numbers. However, one counterexample is sufficient to show that an assertion does not hold generally. For example, the assertion that all natural numbers are less than a trillion passes billions of trial tests but fails for the number one trillion (and all larger integers).

Table 2.1 presents a chart for the sum $S(n) = 1^n + 2^n + 3^n + 4^n$ for some initial values of n. Naturally, it's easy to determine whether or not 5 divides $S(n)$. Just check if the last digit of $S(n)$ is either a 0 or a 5. The assertion is satisfied at least for the first ten values of n. In fact, we might hypothesize an even stronger assertion; namely, that $S(n)$ is divisible by 10 if and only if n is not divisible by 4. Furthermore, when n is divisible by 4 then the last digit of $S(n)$ will always be 4.

TABLE 2.1

n	1	2	3	4	5	6	7	8	9	10
$S(n)$	10	30	100	354	1,300	4,890	18,700	72,354	282,340	1,108,650

For the sake of clarity and uniformity of presentation, now is a good time to introduce the notion of *congruence*. Let a and b be integers and n a natural number. We say that a is *congruent* to b modulo n if n divides $a - b$, written $a \equiv b \pmod{n}$. What a simple idea! We'll see it's also a very compact and convenient way to express a host of arithmetic statements.

Double-check your understanding by verifying that $5 \equiv 17 \pmod{3}$, $32 \equiv 142 \pmod{10}$, and $15 \equiv 3 \pmod{6}$. However, 15 is not congruent to 3 modulo 8, that is, 15 is *incongruent* to 3 modulo 8.

An integer divided by 5 has only one of five possible remainders, namely 0, 1, 2, 3, or 4. We say that the set $\{0, 1, 2, 3, 4\}$ forms a *complete set of residues* modulo 5. Keeping in mind the definition of $S(n)$, let's look at the powers of 1, 2, 3, and 4 (modulo 5). Clearly $1^n = 1$ for all $n \geq 1$, which is congruent to 1 (mod 5). Next $2^n = 2, 4, 8, 16, 32, 64, 128, 256, \ldots$. Reducing mod 5, we obtain $2, 4, 3, 1, 2, 4, 3, 1, \ldots$, which repeats the pattern 2, 4, 3, 1 ad infinitum. We summarize by noting that $2^n \equiv 2, 4, 3$, or $1 \pmod{5}$ depending on whether $n \equiv 1, 2, 3$, or $0 \pmod{4}$, respectively. Similarly, $3^n \equiv 3, 4, 2$, or $1 \pmod{5}$ depending on whether $n \equiv 1, 2, 3$, or $0 \pmod{4}$, respectively. Finally, for $n \equiv 1$, 2, 3, or $0 \pmod{4}$, $4^n \equiv 4, 1, 4, 1 \pmod{5}$.

Now we combine our results by looking at $S(n)$ for each residue modulo 4:

If $n \equiv 1 \pmod{4}$, then $S(n) \equiv 1 + 2 + 3 + 4 = 10 \equiv 0 \pmod{5}$.
If $n \equiv 2 \pmod{4}$, then $S(n) \equiv 1 + 4 + 4 + 1 = 10 \equiv 0 \pmod{5}$.
If $n \equiv 3 \pmod{4}$, then $S(n) \equiv 1 + 3 + 2 + 4 = 10 \equiv 0 \pmod{5}$.
If $n \equiv 0 \pmod{4}$, then $S(n) \equiv 1 + 1 + 1 + 1 = 4$, which is incongruent to 0 (mod 5).

Since all integers are congruent to exactly one of 0, 1, 2, or 3 (mod 4) and only those divisible by 4 are congruent to 0(mod 4), it follows that $1^n + 2^n + 3^n + 4^n$ is divisible by 5 if and only if n is not divisible by 4.

It is no doubt apparent that there is a great deal more structure to these sets of residue classes. The foundations for the development of much of modern number theory and abstract algebra depends on a deep understanding of general principles underlying our discussion of Problem #1. For now, just notice that modular arithmetic is a generalization of "clock arithmetic." The hours of a clock repeat themselves every 12th hour. So telling time is akin to counting modulo 12. Similarly, the days of the week form a residue system modulo 7. Similarly, nearly all cyclic systems having discrete states admit to a modular description. Carl Friedrich Gauss (1777–1855) was the first to state the definition of congruence at the beginning of his magnum opus, *Disquisitiones Arithmeticae* (1801). A great deal of previously described mathematics as well as a host of new discoveries could then be described in terms of the new language of congruences and residue systems. This ability to identify the common thread that unifies several seemingly disparate phenomena is one mark of a great mathematician.

Problem #2 (1983 Green Chicken Contest): *Good news! It is now possible to get Chicken McNuggets in boxes of 6, 9, and 20. What is the smallest integer M such that for any $n \geq M$ it is possible to order exactly n delicious McNuggets by choosing the appropriate number of boxes of each size?*

What a beautiful problem! It may surprise you that there even is such an M. Experiment a bit to find some values of n that work and some values of n that do not admit such an order. For example, can you order exactly 19 McNuggets? How about 33?

The solution reduces to finding the first instance where any number of McNuggets can be ordered for six consecutive numbers; for if we can order any of M, $M + 1$, $M + 2$, $M + 3$, $M + 4$, or $M + 5$ McNuggets, then we can add an appropriate number of boxes of 6 to make any larger order. The key is that any six consecutive numbers cover all residue classes modulo 6. Now which value of M works?

Notice that both 6 and 9 are congruent to 0 (mod 3), while 20 is congruent to 2 (mod 3). So $40 = 20 + 20$ is the smallest number congruent to 1 (mod 3) that can be ordered. This is a promising place to start checking cases. Now $41 = 20 + 9 + 6 + 6$ and $42 = 9 + 9 + 9 + 9 + 6$. Unfortunately, we cannot make an order of size 43 (check it). However, we can make orders for 44, 45, 46, 47, 48, and 49 (please verify). So the smallest integer M such that for any $n \geq M$ it is possible to order exactly n McNuggets is 44. Hmm ... something worth chewing on.

Problem #3 (1984 Green Chicken Contest): *Mr. and Mrs. Gauss invite four other couples to dinner. As the guests arrive they shake hands with everyone they know and no one else (of course they do not shake hands with their spouses or with themselves). As the guests are being seated for dinner Mr. Gauss proclaims, "Not including myself I noticed that one of you shook hands with no one, one of you with just one person, one with two people, ... , and finally one with eight others." How many hands did Mrs. Gauss shake?*

This is another fun problem and at first glance seems impossible to solve. But sit back, relax, listen to Jerry Lee Lewis's "Whole Lotta Shakin' Goin' On", and work a way at the problem one handshake at a time. Please note that although Mr. Gauss does not report on the number of hands he shook, he does include himself when counting the number of hands each guest shook. Let's denote the couples by A and a, B and b, C and c, and D and d. I make no assumptions about which one is the husband or wife. (The problem may seem somewhat dated relative to assumptions about the couples being heterosexual married pairs. But as a math problem, it still works fine.)

First notice that Mrs. Gauss could not possibly have shaken all eight guests' hands. For then, every one of the guests would have shaken at least one hand contrary to the assumption that one person shook no hands. So let A denote the guest who shook eight hands (necessarily including the hands of both Mr. and Mrs. Gauss). Since one person shook no hands, that person must be the spouse of A, namely guest a. Again if Mrs. Gauss shook exactly seven hands, then every guest but a shook at least two hands (person A and Mrs. Gauss), a contradiction. So someone else, say guest B shook exactly seven hands. But then B's spouse, person b, shook exactly one hand. Similarly, one of the guests, say C, shook exactly six hands (and C's partner c shook two hands). In addition, another guest, person D, shook five hands and d shook three hands. This forces Mrs. Gauss to have shaken exactly four hands.

The solution is completely constructive. Figure 2.1 shows precisely who shook whose hands. The amazing thing is that the given information, as inadequate as it seemed at first, allows us complete knowledge of the situation (à la Sherlock Holmes). For example, we now know that Mr. and Mrs. Gauss both know exactly one member of each couple, in fact, the same person in each case.

Problem #4 (1988 Green Chicken Contest): *Show that it is impossible to weight two coins so that the probability of the three outcomes, two heads, a tail and a head, or two tails are all equally likely.*

Before we tackle this problem, a short reminder about basic probability is useful. The likelihood or probability of an event is assigned a number between 0 and 1 inclusive (or 0%–100%). If p is the probability of an event occurring, then $1-p$ is the probability that it did not occur. For example, if the probability that it will rain tomorrow is 40%, then the probability that it will not rain tomorrow is necessarily 60%. If E is an event, the we denote the probability that E occurs by p(E).

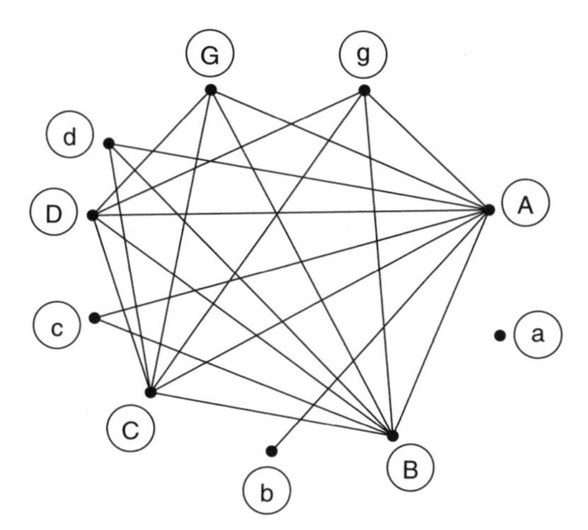

Figure 2.1 Handshaking diagram for five couples.

If A and B are independent events with respective probabilities p_1 and p_2, then the probability that both A and B occur is $p_1 \cdot p_2$. For example, the probability of rolling a 6 with a fair die is 1/6, and the probability of choosing a king by drawing one card from the top of a deck is 1/13 since there are four kings out of 52 total cards. The probability of rolling a 6 and choosing a king is $1/6 \cdot 1/13 = 1/78$ since the two events are independent of each another. But the probability of choosing a king and then choosing another king (without replacing the first card) is not $1/13 \cdot 1/13$ because the first event alters the probability of the second event. Instead, the probability of choosing two kings in succession is $4/52 \cdot 3/51$.

Let's get a feel for the problem by studying the case of two fair coins (where the chance of flipping a head or a tail is equiprobable.) Let H represent the event of getting heads with the first coin and let T represent tails. Similarly for the second coin, let h and t represent the events of heads and tails, respectively. The probability of any single event with either coin is one-half. Since the result of each coin flip is independent of the other coin, the four possible events, *Hh*, *Ht*, *Th*, *Tt*, are all equally likely. In this case the probability of flipping two heads is $p(Hh) = p(H) \cdot p(h) = 1/2 \cdot 1/2 = 1/4$, the probability of flipping a head and a tail is $p(Ht \text{ or } Th) = 1/4 + 1/4 = 1/2$, and the probability of flipping two tails is $p(Tt) = 1/4$. Certainly the three events are not equally likely with two fair coins.

Before we present the solution, it is useful to recall the quadratic formula. If $ax^2 + bx + c = 0$, then $x = \frac{-b \pm \sqrt{b^2 - 4ac}}{2a}$. In particular, if the discriminant $b^2 - 4ac < 0$, then there is no real solution to the quadratic equation (but rather two complex conjugate roots). We are now prepared to solve the two coin problem.

Solution to Problem #4: *To simplify notation, let* $p(H) = P$, $p(T) = Q$, $p(h) = p$, *and* $p(t) = q$. *Of course,* $Q = 1 - P$ *and* $q = 1 - p$, *but we no longer assume that either* p *or* P *equals one-half. Since the events of flipping each coin are independent,* $p(Hh) = Pp$, $p(Ht \text{ or } Th) = Pq + Qp$, *and* $p(Tt) = Qq$. *If the chances of each of these three events is equally likely, each has a probability of one-third of occurring. Observe that* $Pp = Qq$. *Hence* $Pp = (1 - P)(1 - p)$ *implying that* $P + p = 1$. *But* $Pp = \frac{1}{3}$ *and so* $p = \frac{1}{3P}$. *Thus,* $P + 1/(3P) = 1$, *which leads to the quadratic equation* $P^2 - P + 1/3 = 0$.

But in this case the discriminant $b^2 - 4ac = -1/3 < 0$. *Thus there is no real solution to the quadratic equation and hence no way to weight two coins as prescribed.* □

Problem #5 (1993 Green Chicken Contest): *At Middlebury College seven students registered for American history, eight students for British history, and nine students for Chinese history. No student is allowed to take more than one history course at a time. Whenever two students from different classes get together, they decide to drop their current history courses and each add the third. Otherwise there are no adds or drops. Is it possible for all students to end up in the same history course?*

It's fun to play around with the numbers a bit and see what arrangements are possible. For example, if an American history and a British history student get together, they will each drop their history course and add Chinese history, resulting in 6 American history students, 7 British history students, and 11 Chinese history students. If British and American history students continue to talk to one another, eventually 23 students could end up in the Chinese history class with one last student remaining in British history. Fairly quickly one gets the sense that it's not possible to get all the students into the same class. But why?

Solution to Problem #5: *The key is to work modulo 3. Initially there are* $7 \equiv 1 \pmod 3$ *students in American history,* $8 \equiv 2 \pmod 3$ *students in British history, and* $9 \equiv 0 \pmod 3$ *in Chinese history. The numbers constitute a complete residue set modulo 3, the total number of courses involved. If two students add Chinese history for example, the numbers in American, British, and Chinese history become 6, 7, and 11. These numbers still form a complete residue set modulo 3. Analogously, the same happens if two students add either American history or British history. In fact, the numbers always form such a set. At each step of the process, the three enrollment numbers (say* x, y, *and* z*) are incongruent to one another modulo 3. At the next step the numbers are* $x - 1$, $y - 1$, *and* $z + 2$. *Since* x *and* y *are incongruent (mod 3), so are* $x - 1$ *and* $y - 1$. *If* $z + 2 \equiv x - 1 \pmod 3$, *then* $x \equiv z + 3 \equiv z \pmod 3$, *a contradiction. Similarly,* $z + 2$ *is incongruent to* $y - 1 \pmod 3$. *But if all the students ended up in one of the classes, say* $x = 0$, $y = 0$, $z = 24$, *then the numbers no longer form a compete residue system modulo 3. Hence it is impossible for all students to end up in the same history class. (Those who don't study this history problem may be forced to repeat it!)* □

Problem #6 (1986 Green Chicken Contest): *Prove that* $\log_{10} 2$ *is irrational.*

A brief digression about the nature of irrational numbers is in order before we proceed with the solution to Problem #6. *Rational numbers* are those numbers which can be written as *ratios* of natural numbers. For example, 22/7 and 67/101 are rational numbers. The Pythagoreans held the world view that "all is number" and that nature could be completely described in terms of the counting numbers and ratios of them. Apparently, the discovery that some simple real numbers were incommensurable, that is, not the ratio of two integers, came as a shock to them. Today we call such numbers *irrational.*

The classic example of an irrational number is $\sqrt{2}$. Physically it appears as the diagonal of a unit square. If you draw such a square with each side 10 centimeters and measure its diagonal, it appears to be about 14 centimeters. So $\sqrt{2}$ is about $14/10 = 7/5$. But certainly this measurement isn't exact. Maybe $\sqrt{2} = 141/100$ or 14,142/10,000. In other words, is it possible to build a ruler with a finer measure that can give us the exact value? What the Pythagoreans discovered to their dismay is that the answer is "no."

To see why $\sqrt{2}$ is irrational, let us argue by contradiction (reductio ad absurdum). Assume that $\sqrt{2}$ is rational. In that case we can write $\sqrt{2} = a/b$ where a and b are positive integers. If we can write $\sqrt{2}$ at all as a ratio of integers, we can certainly divide out any common divisor and so we may assume in addition that a and b are relatively prime. Now square both sides obtaining $2 = a^2/b^2$. Multiply both sides by b^2 to get $2b^2 = a^2$. Since the left-hand side of the equation is divisible by two, so is the right-hand side. Hence, a^2 is even. But if a were odd, then a^2 would be odd. Thus, a itself is even. Hence there is an integer c for which a = 2c. Substituting 2c for a, we obtain $2\text{b}^2 = (2\text{c})^2$, or equivalently $2b^2 = 4c^2$. Dividing by two leads to $b^2 = 2c^2$. Now the right-hand side of the equation is divisible by two as is the left-hand side. Thus b^2 is even. But if b were odd, then b^2 would be odd. Hence b itself is even. But then both a and b are divisible by two, contradicting the assumption that a and b were relatively prime. It follows, that $\sqrt{2}$ cannot be written as the ratio of integers and hence that $\sqrt{2}$ is irrational.

Before presenting the solution to our last problem, begin by recalling the meaning of $\log_a b$ (the *logarithm base a* of b). For natural numbers a and b, $r = \log_a b$ if $a^r = b$. For example, $\log_{10} 1{,}000 = 3$ since $10^3 = 1{,}000$. Additionally, $\log_9 27 = 3/2$ since $9^{3/2} = 9\sqrt{9} = 27$. The solution to Problem #6 is now fairly straightforward.

Solution to Problem #6: *Assume that $\log_{10} 2$ is rational. Hence there exist relatively prime positive integers a and b for which $\log_{10} 2 = a/b$. By the definition of logarithms, $10^{a/b} = 2$. Next raise both sides of the equation to the bth power. The result is that $10^a = 2^b$. But this leads to $5^a = 2^{b-a}$, which contradicts the Fundamental Theorem of Arithmetic, namely that every integer has a unique prime factorization.* □

The Green Chicken contest is an enjoyable event where each problem is very specific, conceptually distinct from the others, and usually doesn't involve any advanced mathematics—real or complex analysis, topology, or abstract algebra. What it does involve is some cleverness, good problem solving skills, and a willingness to wrestle with a problem for awhile and uncover its underlying structure. Some knowledge of elementary number theory, geometry, and combinatorics often proves helpful. Problems #1 and #5 dealt with congruences (mod 5 and mod 3, respectively). In Problem #5 the underlying structure (of residue systems modulo 3) was not explicit in the statement of the problem. But upon reflection its mathematical structure surfaced. Problem #4 dealt with basic probability in the guise of flipping two coins. The problem can be extended to the situation with three or more coin flips or to rolling a pair of dice, such is the power of mathematics to generalize. Problem #2 was number-theoretic and relates to an area known as Diophantine equations as well as to the concept of covering systems. But I am convinced that anyone willing to think about the problem seriously can solve it. Problem #3 turned out to be a graph theoretic problem. Once we realized that each person could be denoted by a vertex and handshakes by edges of a graph joining vertices, then the solution was immediately amenable to solution. Finally, Problem #6 leads to a deeper understanding of the real number system and the nature of rational and irrational numbers. In all cases, hopefully your interest is piqued to learn more.

In the next chapter, we will deal with one topic and delve a bit wider and deeper. This is more typical of mathematics in general. Mathematicians frequently answer one specific question that in turn creates a host of new, related questions. Oftentimes only once the follow-up questions have been investigated do mathematicians realize the proper setting for the original problem. The answering of interesting questions and the growth of new related areas of investigation is how mathematics develops and renews itself.

WORTH CONSIDERING

1. Show that $n^3/3 + n^2/2 + n/6$ is an integer for all $n \geq 1$.
2. Show that $n^5 - n$ is divisible by 30 for all $n \geq 1$.
3. Show that $S(n) = 1^n + 2^n + 3^n + 4^n$ is divisible by 10 if and only if n is not divisible by 4.
4. Assume that chocolate-covered strawberries come in boxes of 5, 7, and 10. What is the largest number of chocolate-covered strawberries that cannot be ordered exactly?
5. At a party everyone kept track of how many times they shook hands.
   - **(a)** Show that the total number of handshakes was even.
   - **(b)** Show that there were an even number of people who shook hands an odd number of times.
6. Two dice are rolled. Show that it is impossible to weight them so that all possible sums are equally likely.
7. In Problem #5 of Chapter 2, would the conclusion change if initially there were seven students registered for American history, nine students for British history, and nine students for Chinese history? What about 7, 9, and 12 students, respectively?
8. The triplets 3, 7, 11 and 13, 47, 61 are two examples of primes in arithmetic progression, namely $p, p+d, p+2d$. Show that in a prime triplet if $p \neq 3$, then d is divisible by 6, while if $p = 3$, then d is not divisible by 3.
9. **(a)** Prove that $\sqrt{3}$ is irrational
   - **(b)** Why doesn't the same sort of proof work on the number $\sqrt{4}$?
   - **(c)** Prove that $(1 + \sqrt{5})/2$ is irrational.

10. Notice that 2^7 is approximately equal to 5^3. How is this helpful in approximating $\log_{10} 2$?

11. A number is *algebraic* if it is the root of a polynomial with rational coefficients. Show that $\sqrt{2} + \sqrt{3}$ is irrational, but algebraic.

3 The Josephus Problem: Please Choose Me Last

Imagine yourself standing in a circle with 40 others in which every second person in succession remaining is to be killed except for the last one standing. Where would you place yourself so as to survive? No this isn't some reality TV show gone haywire. Rather this is the situation that Josephus Flavius reportedly found himself in the year 66 C.E. when surrounded by a hostile Roman legion. Whether or not Josephus's account of his miraculous survival is completely truthful, we do know that Josephus survived to live an eventful and full life as historian and counselor to several Roman emperors. In addition, his name survives as the progenitor of all related mathematical problems.

There are nearly as many versions of the so-called Josephus Problem as there are people who have written about it. Versions of the mathematical problem of determining where to stand dates back at least to Abraham ibn Ezra (ca. 1092–1167), a prolific Jewish scholar and author of works on astrology, the cabala, mathematics, and philosophy. In later versions the number of people tends to vary from 30 to 41 with every second, third, or even seventh person eliminated. In several accounts, both Josephus and a friend are spared. A medieval version of the problem involves 15 Turks and 15 Christians on board a storm-ridden ship that is certain to sink unless half the passengers are thrown overboard. There's even a Japanese version by Yoshida Koyu appearing in his text, *Treatise on Large and Small Numbers* (1627). This version involves a family of 30 children, half from a former marriage. To choose a child to inherit the parents' estate, they are arranged in a circle with every tenth child eliminated from consideration. The current wife cleverly arranges the children so that none of hers are among the first 15 to be eliminated. However, after 14 children are counted out, the father realizes what is happening and decides to reverse the order and count in a counterclockwise direction. Even so, a child from his second marriage is eventually chosen. I think we can safely assume Father's Day was not a big holiday in that family.

To make sure the situation is clear, consider just seven people with every other one eliminated. Figure 3.1 shows the elimination procedure with person number 7 being the last one to go. The order of elimination is 2, 4, 6, 1, 5, 3, 7.

Mathematical Journeys, by Peter D. Schumer
ISBN 0-471-22066-3

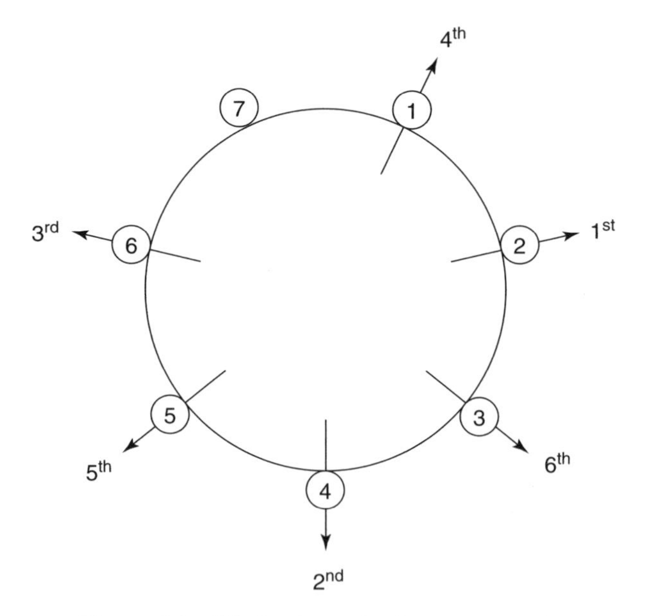

Figure 3.1 Elimination order for seven people.

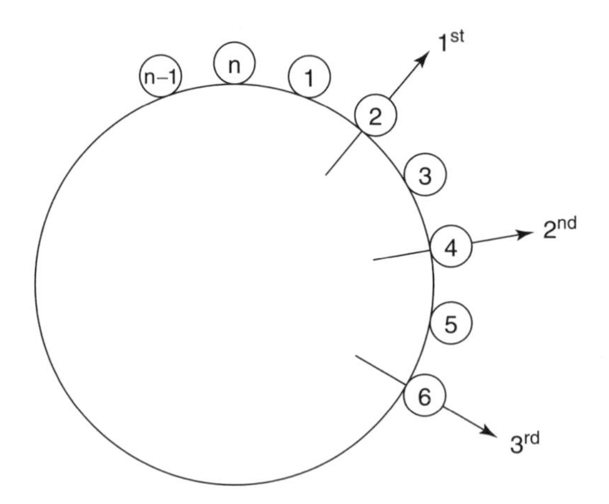

Figure 3.2 Elimination order for n people.

Now let's tackle the following general problem: There are n people numbered one to n standing in a circle. The second person tells a secret to the fourth person, who in turn tells it to the sixth. The process continues as each person tells the secret to the second person ahead clockwise who has not yet been told. Who is the last to know (Fig. 3.2)?

For a given value of n, the problem can of course be solved directly. For example, if $n = 41$, then write out the numbers from 1 to 41 in a circular fashion and cross out every second number until only one number remains. But what if we then wanted to know the answer for 42 people? Since the elimination order depends heavily on the original number in the circle, there doesn't appear to be an easy way to use our first answer to obtain the second. Instead, it appears that we've got to start spinning around all over again. And what about with 43 people—yet another dizzying solution? Looks like a good opportunity for a general mathematical solution.

Let $J(n)$ equal the last person chosen out of a circle of n people when every second person is eliminated. We've already calculated that $J(7) = 7$. Here is a larger chart of $J(n)$ for n from 1 to 20 inclusive (Table 3.1). There seems to be a regular pattern of some sort. Can you accurately describe it? We will in a moment and then we will verify it with a proof. Mathematical induction will play a central role.

Assume $n > 1$ and that all n people are arranged in a circle. There are two cases to consider depending on whether n is even or odd. If n is even (say $n = 2k$), then after the first go-around all the even-numbered people have been removed. We now have a circle with k people and we can pretend that we are just beginning the process starting at person number 1. However, the people are numbered in a funny way. Instead of being numbered from 1 to k consecutively, they are numbered from 1 to $2k-1$ with consecutive odd numbers only (Fig. 3.3a). Hence $J(2k)$, the solution to our problem, is directly related to

TABLE 3.1 Josephus numbers for $1 \leq n \leq 20$

n	1	2	3	4	5	6	7	8	9	10	11	12	13	14	15	16	17	18	19	20
$J(n)$	1	1	3	1	3	5	7	1	3	5	7	9	11	13	15	1	3	5	7	9

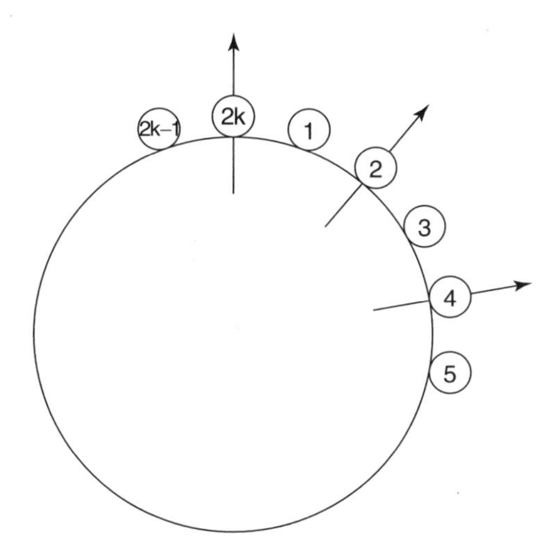

Figure 3.3a Even case: $n = 2k$.

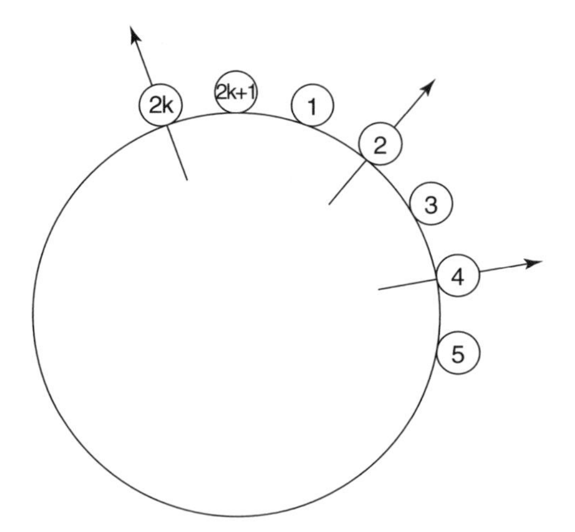

Figure 3.3b Odd case: $n = 2k + 1$.

the solution $J(k)$ with a key modification. Instead of solving for $J(k)$ on the first k natural numbers, we solve for $J(k)$ on the first k odd integers. Thus

$$J(2k) = 2J(k) - 1 \tag{3.1}$$

If n is odd (say $n = 2k + 1$), then after the first go-around again all the even-numbered people have been removed. However, we begin one step back at person number $2k + 1$. The next person removed is person number 1. At this point we continue with a circle of k people numbered $3, 5, \ldots, 2k + 1$ (Fig. 3.3b). We have now reduced the problem of calculating $J(2k + 1)$ to calculating the $J(k)$ problem on the set of odd integers larger than 1. Hence

$$J(2k + 1) = 2J(k) + 1 \tag{3.2}$$

Looking at Table 3.1, it should be evident that $J(n)$ runs through the odd numbers consecutively restarting at 1 each time n is a power of 2. In particular, for $a \geq 0$, let $2^a \leq n < 2^{a+1}$. Hence $n = 2^a + t$ where $0 \leq t \leq 2^a$.

Proposition 3.1:

$$J(2^a + t) = 2t + 1. \tag{3.3}$$

Proof of Proposition 3.1: *Note that $J(1) = 1$ as claimed. Now we will establish the proposition by using induction on the exponent a.*

Let $a = 1$. If $t = 0$, then $n = 2$ and $J(2) = 1$. If $t = 1$, then $n = 3$ and $J(3) = 3$. Hence the proposition is true for $a = 1$.

Now assume that Formula 3.3 holds for some $a - 1$ and all such appropriate values of t. Our inductive step has two parts depending on whether n is even or odd (and hence whether t is even or odd, respectively.)

If n is even, then $n = 2^a + t = 2k$ *for some integer* k*. In this case,*

$$\begin{aligned} J(2^a + t) &= 2J(2^{a-1} + t/2) - 1 \text{ by Formula 3.1} \\ &= 2(2 \cdot \frac{t}{2} + 1) - 1 \\ &= 2t + 1, \text{ as desired.} \end{aligned}$$

If n is odd, then $n = 2^a + t = 2k + 1$ *for some integer* k*. In this case,*

$$\begin{aligned} J(2^a + t) &= 2J(2^{a-1} + \frac{t-1}{2}) + 1 \text{ by Formula 3.2} \\ &= 2(2 \cdot \frac{t-1}{2} + 1) + 1 \\ &= 2t + 1, \text{ as required.} \end{aligned}$$

Thus $J(2^a + t) = 2t + 1$. □

For example, let us calculate $J(100)$. Since 64 is the largest power of two less than or equal to 100, we write $100 = 64 + 36$. By Proposition 3.1, $J(100) = 2 \cdot 36 + 1 = 73$. Of course to get that desired spot, you may have to shove Josephus out of the way.

A useful way to view Proposition 3.1 involves the binary (base 2) representation of numbers. For example, $100 = 1 \cdot 64 + 1 \cdot 32 + 0 \cdot 16 + 0 \cdot 8 + 1 \cdot 4 + 0 \cdot 2 + 0 \cdot 1$ and hence $100_{10} = 1100100_2$. In this case, $t = 100100_2$ and $J(100) = 2t + 1 = 1001001_2$. Note that this is the same as taking the binary representation of 100 and simply removing the leading 1 and then appending it to the right end, formally known as a one-bit cyclic shift left. How simple! It follows that if the binary representation of an integer n consists solely of a string of 1s, then $J(n) = n$. So $J(2^t - 1) = 2^t - 1$ for all $t \geq 1$ since numbers of the form $2^t - 1$ are those with binary representation consisting of all 1's.

Most of us are more comfortable with the decimal (base 10) representation of integers, but the natural language of computers involves binary representation. A simple algorithm to convert the decimal representation of a number into binary is to do the following: write the number n and successively list the result of dividing by 2 (ignoring remainders) and stop when the number 1 is obtained. Now reading from 1 backwards to n, write 1 for each odd number in the list and write 0 for each even number. The result is the binary representation of the original (decimal) number. For example, if $n = 100$, then we write $100 \to 50 \to 25 \to 12 \to 6 \to 3 \to 1$. Reading from right to left we get 1100100, the binary representation of 100.

To take one more example, let us calculate $J(1{,}000)$. The "divide by 2" tabulation becomes $1{,}000 \to 500 \to 250 \to 125 \to 62 \to 31 \to 15 \to 7 \to 3 \to 1$ from which we derive $1{,}000_{10} = 1111101000_2$. Shifting the leading 1 to the right gives us $J(1{,}000) = 1111010001_2 = 977$.

Going the other way is straightforward too. To mentally convert from binary to decimal, start with the number 1 corresponding with the leading 1 on the left of the binary number. Read from left to right. For each 0 double the current number, and for each 1 double the current number and add one. (Equivalently, at each step double and add the current bit.) For example, if we wish to convert the number 110100101_2, we make the following simple calculations: $1 \to 3 \to 6 \to 13 \to 26 \to 52 \to 105 \to 210 \to 421_{10}$. Let's call this the "multiply by 2" procedure as opposed to the "divide by 2" procedure to convert from decimal to binary.

Now let's generalize the elimination process beyond eliminating every other survivor. Define $J(n, q)$ = last person remaining when every qth person is eliminated beginning with n people. Hence $J(n, 2) = J(n)$. Try calculating $J(7, 3)$ for size. Here is a chart of the initial values of J(n, 3) (Table 3.2).

Certainly there is some regularity to the values of $J(n, 3)$, but there does not appear to be a simple, closed formula. It appears that in order to calculate $J(8, 3)$ say, we have to start all over—essentially reinventing the wheel. Even lacking a general formula, it would be extremely useful to be able to use $J(7, 3)$ somehow to calculate $J(8, 3)$. In general, what we seek is a formula for $J(n + 1, q)$ in terms of $J(n, q)$ for any n and q. Here is the result.

Proposition 3.2: *$J(n + 1, q) \equiv J(n, q) + q \pmod{n + 1}$ for $n \geq 1, q \geq 1$.*

Note that we use the numbers $1, 2, \ldots, n + 1$ as our set of complete residues modulo $n + 1$. One consequence of Proposition 3.2 is that $J(n + 1) \equiv J(n) + 2 \pmod{n + 1}$. Double-check this with some examples from Table 3.1.

Another point to realize is that q can be larger than n. Notice that $J(7, 10)$ does not equal $J(7, 3)$ even though $10 \equiv 3 \pmod 7$ and both situations begin with seven people. The order of elimination is different after the first person has been removed. In this case, $J(7, 3) = 4$ while $J(7, 10) = 5$. Check it.

Proof of Proposition 3.2: *Consider $n + 1$ people in a circle with every qth person eliminated. After the first person (number q modulo $n + 1$) is eliminated we now begin the $J(n, q)$ problem, but start q places ahead (modulo $n + 1$). Thus $J(n + 1, q) = J(n, q) + q \pmod{n + 1}$.* □

By Proposition 3.2, $J(8, 3) = J(7, 3) + 3 = 4 + 3 = 7 \pmod 8$. Similarly, $J(9, 3) = J(8, 3) + 3 = 10 \equiv 1 \pmod 9$. Since J($1, q$) = 1 for any value of q, it is now a simple matter to fill in a chart of $J(n, q)$ for any q. All that needs to be done is to repetitively add q, each time reducing modulo the position in the chart. To check your understanding, fill in the following chart for $J(n, 5)$ (Table 3.3).

TABLE 3.2 Josephus numbers with every third person eliminated

n	1	2	3	4	5	6	7	8	9	10	11	12	13	14	15	16	17	18	19	20	21	22	23	24
$J(n, 3)$	1	2	2	1	4	1	4	7	1	4	7	10	13	2	5	8	11	14	17	20	2	5	8	11

TABLE 3.3 Josephus numbers with every fifth person eliminated

n	1	2	3	4	5	6	7	8	9	10	11	12	13	14	15	16	17	18	19	20
$J(n, 5)$	1	2	1	2	2	1	6	3	—	3	—	—	6	11	—	6	—	—	—	7

It is often convenient to list the actual order of elimination. We call this the *Josephus permutation* and denote it in the following way: $P(n, q) = \begin{pmatrix} 1 & 2 & 3 & \dots & n \\ r_1 & r_2 & r_3 & \dots & r_n \end{pmatrix}$. Here r_1 is the first person removed, r_2 is the second person removed, and so on. Hence r_n is necessarily $J(n, q)$. The Josephus permutation gives us the full story of the elimination procedure, not just the position of the last person remaining. For example, $P(7, 3) = \begin{pmatrix} 1 & 2 & 3 & 4 & 5 & 6 & 7 \\ 3 & 6 & 2 & 7 & 5 & 1 & 4 \end{pmatrix}$. Another way to view this is to think of the following card shuffle: Place n cards numbered 1 through n and flip through them placing every qth card facedown on a table. The resulting order of the cards is the same as the Josephus permutation. (We can call this a Josephus shuffle—certain to impress our friends at the next poker night.)

There is a host of interesting questions to be asked about the order of elimination. For example, given n people can we eliminate them in their original order $1, 2, 3, \dots, n$ by choosing an appropriate value of n? Of course we can by picking $q = 1$. Is there a larger value of q that does the trick? Again the answer is yes; let $q = n! + 1$. Let's see why.

In general, adding $n!$ to a given value of q will not alter the Josephus permutation since every natural number less than or equal to n divides evenly into $n!$. At any stage of the elimination with r people remaining, instead of moving q steps we spin around $n!/r$ times and then move q steps. But we always land at the same place and so $P(n, q) = P(n, q + n!)$. In fact, by similar reasoning if we let $L(n)$ be the least common multiple of the numbers $1, 2, \dots, n$, then $P(n, q) = P(n, q + L(n))$. For example, $L(7) =$ least common multiple of $1, 2, 3, 4, 5, 6, 7 = 2^2 \cdot 3 \cdot 5 \cdot 7 = 420$. So $P(7, 3) = P(7, 423)$.

Given n people can we choose q so that the n people are eliminated in reverse order? Our discussion above holds the key. Let $q = L(n)$. Then the Josephus process repeatedly picks off the last person left in the circle as desired. For example, $P(7, 420) = \begin{pmatrix} 1 & 2 & 3 & 4 & 5 & 6 & 7 \\ 7 & 6 & 5 & 4 & 3 & 2 & 1 \end{pmatrix}$.

Now let us assume that n is an even number. The solution above solves the problem of eliminating the second half of the circle first for any even n, namely let $q = L(n)$. With n an even number and $q = 2$ the Josephus process eliminates all the even-numbered positions first, in fact in numerical order.

What if we want to eliminate all the odd-numbered positions first? $P(4, 5) = \begin{pmatrix} 1 & 2 & 3 & 4 \\ 1 & 3 & 4 & 2 \end{pmatrix}$. But then we hit a snag.

Given $n = 6$, no value of q will first eliminate positions 1, 3, 5 in that precise order. If there were such a value of q, then $q \equiv 1 \pmod 6$ in order to eliminate person #1 first. But now there are 5 people left and beginning at position 1 we wish to eliminate person #3. Since person #2 also remains, it follows that $q \equiv 2 \pmod 5$. Next there are 4 people remaining, we start at position 3, and must eliminate person # 5 stepping over person #4. Thus $q \equiv 2 \pmod 4$. But it cannot be the case that q is both congruent 1 (mod 6) and 2 (mod 4). The first congruence implies that q is odd; the second congruence says that q is even. Looks like we're beat. So let's relax our expectations a bit and see if we can choose a value of q that eliminates the odd positions first, but not require that they be in ascending order.

Interestingly, we can always solve the above problem eliminating the odd-numbered positions first all done in descending order! Just let $q = L(n) - 1$ and the numbers fall neatly into place. By way of example, consider $n = 6$. In this case, $L(6) = 60$ and $L(6) - 1 = 59$. We obtain $P(6, 59) = \begin{pmatrix} 1 & 2 & 3 & 4 & 5 & 6 \\ 5 & 3 & 1 & 4 & 6 & 2 \end{pmatrix}$. This solves our problem nicely, but it is of interest to note that $L(n) - 1$ doesn't necessarily provide the smallest value of q, which eliminates odd positions first. In the example with $n = 6$, it so happens that $P(6, 19) = \begin{pmatrix} 1 & 2 & 3 & 4 & 5 & 6 \\ 1 & 5 & 3 & 4 & 6 & 2 \end{pmatrix}$. Again all the odd numbers are knocked out before any of the even numbers, but not in strict descending order. Finding the smallest value of q that eliminates all odds first is still an unsolved problem. Care to try?

WORTH CONSIDERING

1. Solve the original Josephus problem mentioned at the beginning of this chapter. Namely, where should you stand to be the sole survivor in a circle of 41 people with every second person killed?
2. In the Japanese version of the Josephus problem with 15 children from each of two marriages, how were the children distributed around the circle. Which child was chosen heir?
3. Find $J(n)$ for $n = 50$, $n = 199$, $n = 512$, and $n = 1{,}000{,}000$.
4. Make a table like Table 3.2 for $J(n, 4)$ for $1 \leq n \leq 25$.
5. Convert the following decimal numbers to binary using the "divide by two procedure": 200, 356, 10,000.
6. Convert the following binary numbers to decimal using the "multiply by two procedure": 1001001, 10101110000, 11011110000.
7. Calculate $P(8, q)$ for values of q from 2 to 12. Notice how many fixed points each permutation has.
8. A fixed point in a shuffle is a card which remains in its original position. In a random Josephus shuffle (or permutation), show that we expect just one fixed point.
9. Complete Table 3.3.

4 Nim and Wythoff's Game: Or How to Get Others to Pay Your Bar Bill

If you plan on going to a bar and hope to have someone else pick up the tab, it's probably best to go with a good friend who already owes you money. But if you hope to win a bet against a stranger, in this chapter we present two related games that just might come in handy. Both games are fun and interesting. The first game is Nim, which was introduced together with a completely worked out strategy by Harvard mathematician, C.L. Bouton in a 1901–1902 *Annals of Mathematics* article. In 1907, a similar game with an additional option was discussed by the Dutch mathematician W.A. Wythoff. Both games are easy to learn and can be enjoyably played by children. But winning strategies take some know-how, which we present below. Let's do the math.

Nim is a two-person game played with identical counters laid out in several piles. The number of counters in each pile need not be the same. Play alternates, and on a given turn a player may remove any positive number of counters from any single pile. The winner is the player who removes the last counter. That's all the rules!

Let's consider an example. Suppose the game begins with three piles consisting of 3, 5, and 7 counters, respectively (Fig. 4.1) and two players, Amy and Bob. Amy removes 3 counters from the middle pile leaving 3, 2, and 7 counters. Bob then removes 6 counters from the right resulting in 3, 2, and 1 counters remaining. Amy removes 1 from the left, leaving piles of 2, 2, and 1. Bob removes 1 from the right pile leaving two piles of 2 apiece. At this point he knows he can win by mimicking Amy's play. If Amy removes either 1 or 2 counters from one pile, Bob will do likewise with the other pile and force a win. Amy loses the game but refuses to pay the bill anyway, a sneaky move Bob unfortunately didn't fully consider. Oh well, back to a strategy for Nim.

A winning strategy in Nim requires making moves that eventually result in victory. If there are n piles, every position can be represented by an n-tuple of nonnegative integers. For example, the sequence of positions in our previous example are represented by $(3, 5, 7) \rightarrow (3, 2, 7) \rightarrow (3, 2, 1) \rightarrow (2, 2, 1) \rightarrow$

Mathematical Journeys, by Peter D. Schumer
ISBN 0-471-22066-3

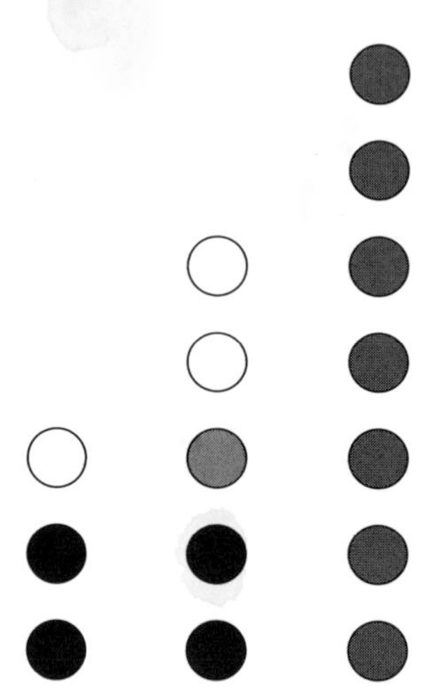

Figure 4.1 Amy's moves in white, Bob's moves in gray, black counters remaining.

$(2, 2, 0)$, etc. Notice that the order of the numbers in parentheses doesn't really matter, hence these are unordered n-tuples. A player has a *winning position* if no matter what move the opponent makes, the player can eventually force a win (through proper play). In other words, a winning position is one that forces the opponent to choose a *losing position*.

Interestingly, developing a best strategy for the game of Nim requires binary representations. Write the number of counters in each pile in binary. Then list the numbers vertically with the units place aligned. The number of 1's in any vertical column will be either even or odd (called an even or odd column, respectively). The player who removes the last counter leaves all piles with zero counters, resulting in all columns being even. In fact, the winning strategy is just that. Winning positions are exactly those that have all even columns. Here's how it works.

At any point in the game, if your opponent leaves a position with at least one odd column, you can remove an appropriate number of counters from one pile that results in all even columns. To accomplish this, go to a pile that contributes a 1 to the left-most odd column. From that pile remove a number of counters that changes all odd columns to even columns. If a column is odd, make it even by changing a 0 to 1 or vice versa. Leave all even columns alone. This is guaranteed to be possible since the new number in the chosen pile becomes smaller (the key column reduces from 1 to 0 and any other column changes occur to the right).

Here's an example. Suppose the game begins with three piles of 29, 23, and 3 counters and we get to play first. Written in binary the numbers become

$$\begin{matrix} 1 & 1 & 1 & 0 & 1 \\ 1 & 0 & 1 & 1 & 1 \\ 0 & 0 & 0 & 1 & 1. \end{matrix}$$

We have appended three 0's to the last number just to make the columns clearer. Reading from left to right the columns are EOEEO where E represent "even" and O represent "odd." The second column is the first odd column and hence we go to a pile contributing a 1 to that column. In this case, the only candidate is the first pile. In order to leave all even columns we must leave 10100 counters in the first pile. So we must leave 20 counters in the first pile (or equivalently remove 9 counters). The result is tabulated as follows:

$$
\begin{array}{ccccc}
1 & 0 & 1 & 0 & 0 \\
1 & 0 & 1 & 1 & 1 \\
0 & 0 & 0 & 1 & 1.
\end{array}
$$

Suppose our opponent removes 10 counters from the second pile leaving this result:

$$
\begin{array}{ccccc}
1 & 0 & 1 & 0 & 0 \\
0 & 1 & 1 & 0 & 1 \\
0 & 0 & 0 & 1 & 1.
\end{array}
$$

Now the first column has an odd number of 1's and so we must move to remedy this. The first pile has a 1 in the first column and hence we will remove some counters from the first pile. In fact, we need to leave 01110 counters, that is, leave 14 counters by removing 6 of them. The resulting piles are

$$
\begin{array}{ccccc}
0 & 1 & 1 & 1 & 0 \\
0 & 1 & 1 & 0 & 1 \\
0 & 0 & 0 & 1 & 1.
\end{array}
$$

From this point on you should be able to work out a winning combination for all subsequent moves by your opponent.

The second game that we consider is Wythoff's game. Again, two players alternate in removing counters and the player to remove the last counter is declared the winner. In this instance there are only two piles of counters, but there is one additional move. At each stage of the game, instead of removing any number of counters from one pile, a player may choose to remove an equal number of counters from both piles. This one additional possibility significantly alters an appropriate winning strategy.

Here's a brief example of a game played between Andrew and Bertha. The two piles begin with 10 and 20 counters, respectively, which we denote by (10, 20). Andrew removes 6 counters from the second pile, hence (10, 14) remains. Next Bertha takes 8 counters from both piles, leaving (2, 6). The following moves are Andrew (2, 5), Bertha (2, 1), Andrew (1, 1), then Bertha (0, 0) and the victory.

Some mathematical background is required before a winning strategy for Wythoff's game will be presented. We begin with some number play. Somewhat arbitrarily, let $x = \sqrt{2}$ and let $y = 1/x$. Next we tabulate $n(1 + x)$ and $n(1 + y)$ to four decimal digits for $1 \leq n \leq 10$ (Table 4.1).

Table 4.1 probably appears to be a fairly random list of real numbers. However, if we modify it slightly, then a great deal of regularity will appear. Instead

TABLE 4.1

n	1	2	3	4	5	6	7	8	9	10
$n(1+x)$	2.4142	4.8284	7.2426	9.6568	12.0710	14.4852	16.8994	19.3137	21.7279	24.1421
$n(1+y)$	1.7071	3.4142	5.1213	6.8284	8.5355	10.2426	11.9497	13.6568	15.3639	17.0710

TABLE 4.2

n	1	2	3	4	5	6	7	8	9	10
$[n(1+x)]$	2	4	7	9	12	14	16	19	21	24
$[n(1+y)]$	1	3	5	6	8	10	11	13	15	17

TABLE 4.3

n	1	2	3	4	5	6	7	8	9	10
$[n(1+x)]$	3	6	9	12	15	18	21	25	28	31
$[n(1+y)]$	1	2	4	5	7	8	10	11	13	14

of calculating $n(1+x)$ and $n(1+y)$, let us calculate $[n(1+x)]$ and $[n(1+y)]$ where $[r]$ denotes the greatest integer less than or equal to r (Table 4.2).

Do you see a peculiar phenomenon? Notice that among the bottom two rows every natural number from 1 to 17 occurs exactly once in addition to a few larger numbers. We may well wonder if this pattern continues for larger values of n. Namely, will all natural numbers eventually appear, and if so, will each appear precisely once? In addition, is this a special property of the number $x = \sqrt{2}$, or does this hold more generally?

Let's make similar calculations for $x = \pi - 1$ and $y = 1/x$ (Table 4.3). Miraculously all small natural numbers seem to appear and to appear just once. In this case, every integer from 1 to 15 appears along with some larger values. Let's look at one more example and then state the result.

Let $x = 4$, $y = 1/x$, and tabulate $[n(1+x)]$ and $[n(1+y)]$. We obtain Table 4.4. This time things did not work out as cleanly. The numbers 4 and 9 do not appear at all, while 5 and 10 appear twice. So this phenomenon seems to be limited to certain real numbers. Here is the result:

Theorem 4.1: *Let $x > 0$ be an irrational number and $y = 1/x$. Consider the sequence $S = \{[1+x], [2(1+x)], [3(1+x)], \ldots, [1+y], [2(1+y)], [3(1+y)], \ldots\}$. Then S contains every positive integer exactly once.*

TABLE 4.4

n	1	2	3	4	5	6	7	8	9	10
$[n(1+x)]$	5	10	15	20	25	30	35	40	45	50
$[n(1+y)]$	1	2	3	5	6	7	8	10	11	12

Theorem 4.1 was first proved by the Canadian mathematician Sam Beatty (1881–1970) in 1926. In fact, we now call such sequences *Beatty sequences* in his honor. More specifically, the sequences $\{[n(1+x)]\}_{n=1}^{\infty}\{[n(1+y)]\}_{n=1}^{\infty}$ are called *complementary Beatty sequences*. Beatty was the recipient of the first Canadian Ph.D. in mathematics (1915) under the direction of J.C. Fields (1863–1932). In fact, Beatty was Fields's only doctoral student. Beatty went on to lead an illustrious career at the University of Toronto where as head of the mathematics department he hired some of the greatest Canadian and international mathematicians including algebraist Richard Brauer, geometer H.S.M. Coxeter, and logician Abraham Robinson.

Beatty's advisor, John Charles Fields, was the creator of the International Medals for Outstanding Discoveries in Mathematics, better known as the Fields Medals. Fields was professor at University of Toronto and was the president of the International Congress of Mathematicians that was held there in 1924. Surplus from monies raised for that congress created the original funding for the illustrious Fields Medals. Later awards have been funded from his estate.

Proof of Theorem 4.1: *Every number of the form $n(1+x)$ is irrational. Otherwise $n(1+x) = p/q$ for integers p and q implying that $x = (p - nq)/nq$, a rational number contrary to our assumption about x. Similarly $n(1+y)$ is irrational for all $n \geq 1$. Hence no member of the sequence S is an integer.*

The number of multiples of $1+x$ that are less than n is exactly $[\frac{n}{1+x}]$. Similarly, the number of multiples of $1+y$ that are less than n is exactly $[\frac{n}{1+y}]$. Now fix n, for some positive integer n. By the definition of the greatest integer function, we have

$$\frac{n}{1+x} - 1 < \left[\frac{n}{1+x}\right] < \frac{n}{1+x} \text{ and}$$
$$\frac{n}{1+y} - 1 < \left[\frac{n}{1+y}\right] < \frac{n}{1+y}.$$

Adding the inequalities together we obtain

$$n\left(\frac{1}{1+x} + \frac{1}{1+y}\right) - 2 < \left[\frac{n}{1+x}\right] + \left[\frac{n}{1+y}\right] < n\left(\frac{1}{1+x} + \frac{1}{1+y}\right).$$

But

$$\frac{1}{1+x} + \frac{1}{1+y} = \frac{1}{1+x} + \frac{x}{1+x} = 1.$$

So

$$n - 2 < \left[\frac{n}{1+x}\right] + \left[\frac{n}{1+y}\right] < n.$$

But $[\frac{n}{1+x}] + [\frac{n}{1+y}]$ is an integer. Hence $[\frac{n}{1+x}] + [\frac{n}{1+y}] = n - 1$.

Thus there are $n - 1$ elements of S below n. Analogously, there are n elements of S below $n + 1$ (by the same analysis with n replaced by $n + 1$). So there is precisely one element of S equal to n. Since n is arbitrary, there is precisely one element of S equal to any given natural number. □

Returning to Wythoff's game, we call (a, b) a *winning position* for player A to create if there is a strategy from (a, b) such that no matter what player B does, A can eventually win the game. Wythoff's amazing theorem (1907) is the following:

Theorem 4.2: *The winning positions in Wythoff's game are (0, 0) and* $([n(1 + x))], [n(1 + y)])$ *for* $n \geq 1$ *where* $x = \frac{-1+\sqrt{5}}{2}$ *and* $y = \frac{1}{x}$.

Note that x is the *golden mean* (or golden section), that ubiquitous number studied by the ancient Greeks. In particular, Chapter II, Proposition 11 of Euclid's *Elements* directs us "to cut a given straight line so that the rectangle contained by the whole and one of the segments is equal to the square on the remaining segment." In modern parlance, we are asked to find a point x along a unit line segment such that the ratio 1 to x is as x to $1 - x$. Solving for x leads to the quadratic equation $x^2 + x - 1 = 0$ from which we obtain the positive solution $x = \frac{-1+\sqrt{5}}{2}$. Study of the golden mean and its natural occurrence within a regular pentagram as well as its apparent aesthetic appeal no doubt dates back to the Pythagoreans, several centuries before Euclid (Fig. 4.2).

Let's make a table for $[n(1+x)]$ and $[n(1+y)]$ for $x = \frac{-1+\sqrt{5}}{2}$ and $y = 1/x$ (Table 4.5). Notice that in every column the difference between $[n(1+x)]$ and $[n(1+y)]$ is exactly n itself. This special property of $x = \frac{-1+\sqrt{5}}{2}$ allows for a winning strategy in Wythoff's game. Any nonwinning position can be changed in one move to a winning position and any move from a winning position will result in a nonwinning position. Next we give a more formal proof. For the sake

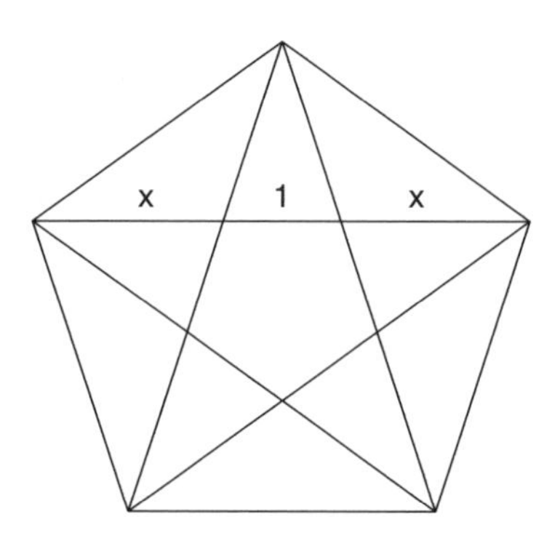

Figure 4.2 Golden mean $x = 1 + 5/2$.

TABLE 4.5

n	1	2	3	4	5	6	7	8	9	10	11	12	13	14	15
$[n(1+x)]$	1	3	4	6	8	9	11	12	14	16	17	19	21	22	24
$[n(1+y)]$	2	5	7	10	13	15	18	20	23	26	28	31	34	36	39

of definiteness, we refer to the two players as Player 1 and Player 2 and each turn as a move. In addition, the two piles will be referred to as pile A and pile B.

Proof of Theorem 4.2: *The number* $x = \frac{-1+\sqrt{5}}{2}$ *is irrational and hence, by Theorem 4.1, the sequences* $\{[n(1+x)]\}_{n=1}^{\infty}$ *and* $\{[n(1+y)]\}_{n=1}^{\infty}$ *form complementary Beatty sequences with every natural number appearing uniquely. In this instance, since* $x - 1/x = 1$, *we have for every* $n \geq 1 : [n(1+y)] - [n(1+x)] = n$.

Assume that at some point Player 1 makes a move resulting in position (a, b) *where the ordered pair* (a, b) *occurs among the entry rows of Table 4.5 and* $a \leq b$. *In particular, if* $b - a = n$, *then* $a = [n(1+x)]$ *and* $b = [n(1+y)]$. *Player 2 has three choices for a move depending on whether counters are removed from (a) pile A, (b) pile B, or (c) both piles. In no case will the new position be among those in Table 4.5 by the uniqueness of its entries (case a and b) and the fact that all natural numbers occur exactly once as the difference between piles (case c).*

In case (a), Player 2 moves to position $(a - s, b)$ *for some* s *with* $1 \leq s \leq a$. *Player 1 can return to a position listed in Table 4.5 by finding the entry* $a - s$ *in the table and moving left to that position at either* $(a^*, a - s)$ *or* $(a - s, b^*)$ *for appropriate* a^* *or* b^*. *In case (b), Player 2 moves to position* $(a, b - s)$ *for some* s *with* $1 \leq s \leq b$. *If* $a \leq b - s$, *then Player 1 removes enough counters from both counters to move* s *places to the left in Table 4.5. There the difference between the two piles is* $b - s - a$ *as required. If* $0 \leq b - s < a$, *then move to the position where* $b - s$ *occurs in the table ending up at either* $(a^*, b - s)$ *or* $(b - s, b^*)$ *for appropriate* a^* *or* b^*. *In case (c), Player 2 moves to position* $(a - s, b - s)$ *for some* s *with* $1 \leq s \leq \min\{a, b\}$. *In this situation, Player 1 finds* $a - s$ *in Table 4.5 and moves accordingly to either* $(a^*, a - s)$ *or* $(a - s, b^*)$ *for some* a^* *or* b^*. *In all three cases, Player 1 moves to a winning position from which Player 2 is forced to move away.* □

Consider the following example: Suppose Amy leaves the position (9, 15) for Andrew—namely two piles with 9 and 15 counters, respectively. Andrew has three possibilities. He can remove any number of counters from the 9's pile, he can remove any number from the 15's pile, or he can remove an equal number from both piles. Suppose he removes four counters from the 15's pile, leaving Amy with position (9, 11). This is case (b), where the difference is $11 - 9 = 2$. So Amy moves to the second position (3, 5) by removing six counters from each pile. Suppose Andrew now removes one counter from the 3's pile, resulting in (2, 5). This is case (a) and so Amy looks for the number 2 in the table. She moves to position (1, 2) by removing four counters from the 5's pile. Now Andrew

TABLE 4.6

n	1	2	3	4	5	6	7	8	9	10	11
A	1	100	101	1001	10000	10001	10100	10101	100001	100100	100101
B	10	1000	1010	10010	100000	100010	101000	101010	1000010	1001000	1001010

must move to (1, 1), (0, 2), or (0,1) from which Amy will remove the remaining counters and win the game.

There is a wonderful way to view the winning positions in Wythoff's game. Use Fibonacci numbers! In particular, write the numbers in pile A using their Zeckendorf representation. Then the corresponding numbers in pile B have the same Zeckendorf representation with the "Fibonacci digit" (or fidget) 0 appended on the right. Table 4.6 is part of Table 4.5 rewritten with integers given as Zeckendorf representations.

You are now fully prepared to step into any local watering hole, walk up to the biggest and meanest customer, and say, "Choose your weapon—Nim or Wythoff's game?"

WORTH CONSIDERING

1. What move(s) creates a winning position in Nim if there are four piles with 5, 8, 9, and 11 counters, respectively?
2. What move in Wythoff's game creates a winning position if the piles have 20 and 40 counters, respectively?
3. Show that in the game of Nim, every move from a winning position results in a losing position.
4. Show that at any step, the number of ways to reach a winning position in the game of Nim is at most the number of nonempty piles.
5. Develop a winning strategy for the following game introduced by C.G. Bachet (1612): Two players add from 1 to 10 counters to a common pile. The player who adds the 100th counter is declared the winner.
6. Northcott's game is played on a checkerboard. Red begins with eight checkers along the first rank and black with eight checkers along the eighth rank. Play proceeds by each player moving one checker any number of spaces along a file toward an opposing checker. The last player able to move wins. Relate Northcott's game to Nim and then develop a winning strategy.
7. **(a)** In the game of One Pile, players alternate removing from 1 to m counters from a common pile. The winner is the player who removes the last counter. Develop a winning strategy for One Pile.

   (b) In the misère version of One Pile, the loser is the one who removes the last counter. How does this change the winning strategy?

8. The game of Thirty-one is played with twenty-four cards, the A, 2, 3, 4, 5, and 6 of each suit representing values one to six. Players alternate by laying down any remaining card on a common sum. The winner is the player who makes the total 31 or who forces the opponent to exceed 31. Investigate the game.

5 Mersenne Primes, Perfect Numbers, and Amicable Pairs

Pythagoras and his followers believed that a disciplined study of mathematics and philosophy was essential to a just and moral life. In fact, it is believed that the very words *philosophy* (meaning love of wisdom) and *mathematics* (meaning that which is learned) may have been coined by Pythagoras himself. The Pythagoreans classified the natural numbers into various categories and attributed religious and mystical meanings to them. For example, one was the generator of all numbers and the source of reason. Even numbers were female and odd numbers beginning with three were male. Four was the number for justice and squaring of accounts. Five, being the sum of the first female and male numbers, represented marriage. Our starting point for this chapter is the number six.

Six was the number of creation that interestingly meshes perfectly well with the Judeo-Christian account of the genesis of the universe. But six was also *perfect* in that it was the sum of its aliquot parts, that is, six is the sum of its proper divisors: $6 = 1+2+3$. Another such example is $28 = 1+2+4+7+14$. Numbers for which the sum of proper divisors was less than the original number were called *deficient*. For example, all primes are deficient as is the number 8, whose proper divisors sum to seven. Finally, if the sum of proper divisors exceeded the number itself, then the number was *abundant*. The number 12 is abundant since $1+2+3+4+6 > 12$.

The ancient Greeks discovered two additional perfect numbers: 496 and 8,128. But that is where their catalog of perfect numbers ends. Even so, they knew that the numbers were far from random and discovered a fair amount about the form of such numbers. Let us retrace their discoveries by factoring the first four perfect numbers:

$$\begin{aligned} 6 &= 2 \cdot 3 \\ 28 &= 4 \cdot 7 \\ 496 &= 16 \cdot 31 \\ 8{,}128 &= 64 \cdot 127. \end{aligned}$$

In each case, we have written the given number as a product of an even and an odd number, a unique and appropriate decomposition from a Pythagorean

Mathematical Journeys, by Peter D. Schumer
ISBN 0-471-22066-3

perspective. Next notice that for each perfect number, its largest odd factor is one shy of being a power of two itself. Furthermore, that power of two is exactly twice as big as the largest even factor of the number. Rewritten the list looks like this:

$$\begin{aligned} 6 &= 2^1 \cdot (2^2 - 1) \\ 28 &= 2^2 \cdot (2^3 - 1) \\ 496 &= 2^4 \cdot (2^5 - 1) \\ 8{,}128 &= 2^6 \cdot (2^7 - 1). \end{aligned}$$

Notice that the powers of two comprising the odd part are all prime numbers, in fact the first four primes! Furthermore, the numbers 3, 7, 31, and 127 are themselves prime. This observation is summarized in Euclid's *Elements*, Book IX, Proposition 36:

Theorem 5.1: *If $2^p - 1$ is prime, then $n = 2^{p-1}(2^p - 1)$ is perfect.*

Before we can present Euclid's proof of this assertion, further background is needed.

Observation 1: *If $2^n - 1$ is prime, then n itself is prime.*

If n were composite, then n could be expressed as $n = a \cdot b$ where $a > 1$ and $b > 1$. But then

$$\begin{aligned} 2^n - 1 &= 2^{ab} - 1 \\ &= (2^a - 1)(2^{a(b-1)} + 2^{a(b-2)} + \ldots + 2^a + 1), \end{aligned}$$

the last product consisting of two numbers larger than 1. Hence if n is composite, $2^n - 1$ is composite. (See the "Worth Considering" section at the end of this chapter for a similar observation about $a^n - 1$ for $a > 2$.)

By the way, the mathematics of geometric sums such as that which appears here was presented in Book IX, Proposition 35 of the Elements.

Define *the sum of divisors function*, $\sigma(n)$, to be the sum of all the positive divisors of n. So $\sigma(6) = 1+2+3+6 = 12$, $\sigma(10) = 1+2+5+10 = 18$, and $\sigma(28) = 56$. For any prime number p, $\sigma(p) = p + 1$. A key observation is that the number n is a perfect number if $\sigma(n) = 2n$. We need to know a little bit more about the sum of divisors function.

Observation 2: *If p is prime and a is a positive integer, then $\sigma(p^a) = (p^{a+1} - 1)/(p - 1)$.*

For verification, let $S = \sigma(p^a)$. Then

$$\begin{aligned} S &= 1 + p + p^2 + \ldots + p^a \text{ and} \\ pS &= p + p^2 + \ldots + p^a + p^{a+1}. \end{aligned}$$

Subtracting S from pS:

$$(p-1)S = p^{a+1} - 1.$$

Hence

$$S = (p^{a+1} - 1)/(p - 1).$$

In particular, if $p = 2$*, then* $\sigma(2^a) = 2^{a+1} - 1$.

The other observation, which we won't formally prove here, is that the sum of divisors function is a *multiplicative* function. Namely, if a and b are relatively prime, then $\sigma(ab) = \sigma(a) \cdot \sigma(b)$. For example, since $\sigma(10) = 18, \sigma(13) = 14$, and 10 and 13 are relatively prime, it follows that $\sigma(130) = \sigma(10 \cdot 13) = \sigma(10)\sigma(13) = 18 \cdot 14 = 252$. The main idea is that if a and b are relatively prime, then the divisors of a and the divisors of b are distinct (other than the number one). Hence the product of any divisor of a with a divisor of b will produce a new divisor of ab, and conversely, all divisors of ab are created in this way.

Proof of Theorem 5.1: *Let* $2^p - 1$ *be prime and set* $n = 2^{p-1}(2^p - 1)$*. Since* 2^{p-1} *and* $2^p - 1$ *are relatively prime, it follows that*

$$\sigma(n) = \sigma(2^{p-1}(2^p - 1)) = \sigma(2^{p-1}) \cdot \sigma(2^p - 1).$$

But by Observation 2, $\sigma(2^{p-1}) = 2^p - 1$*. Furthermore, since* $2^p - 1$ *is prime,* $\sigma(2^p - 1) = (2^p - 1) + 1 = 2^p$*. Hence*

$$\sigma(n) = (2^p - 1)2^p = 2 \cdot 2^{p-1}(2^p - 1) = 2n. \qquad \square$$

It follows from Theorem 5.1 that every discovery of a prime of the form $2^p - 1$ leads to a concomitant perfect number. During the Middle Ages three more primes of this form were discovered for prime values $p = 13$, 17, and 19. In the preface of his book, *Cogita Physica-Mathematica* (1644), the French Minimite friar Marin Mersenne (1588–1648) claimed that $2^p - 1$ is prime for the following values of p: 2, 3, 5, 7, 13, 17, 19, 31, 67, 127, and 257 and for no other values of $p \leq 257$. Mersenne was a noted scholar and maintained a voluminous correspondence with the great European mathematicians of his day. His proclamation about the primality of $2^p - 1$ led to a reawakening of interest in such primes and the related issue of perfect numbers. In his honor, today we call such primes *Mersenne primes*.

The first significant step toward either verifying or debunking Mersenne's conjecture was taken by the incomparable Swiss mathematician Leonhard Euler (1707–1783). Euler made use of a result of Pierre de Fermat's (c. 1601–1665), which stated that all factors of $2^p - 1$ are of the form $2np + 1$ for some n. Hence Euler reduced the problem of checking the primality of $2^{31} - 1$ by checking trial divisors of the form $62n + 1$. In fact, Euler just needed to check for primes of that form up to $\sqrt{2^{31}-1}$, approximately 46,341. Of course, having Euler's rapid arithmetic capabilities was especially helpful as well. Since none of the trial divisors turned out to be a legitimate divisor, Euler verified (1772) that in fact, $2^{31} - 1 = 214{,}7483{,}647$ is prime. This was an impressive feat at the time. In Peter Barlow's *Theory of Numbers* (1811), he states that this prime number "is the greatest that will ever be discovered, for, as they are merely curious without being useful, it is not likely that any person will attempt to find one beyond it."

Leonhard Euler

Subsequent mathematical history has not been kind to either Peter Barlow's or to Mersenne's conjecture. For over a hundred years, Euler's prime was the largest prime known. But in 1876, Edouard Lucas (1842–1891) developed a new primality test for Mersenne primes and was able to verify that $2^{127} - 1$ was prime. Good news for Mersenne; bad news for Barlow. But in 1903, Frank Nelson Cole (1861–1926) gave a silent albeit stunning presentation at the American Mathematical Meeting in New York City by calculating $2^{67} - 1$ and then demonstrating that it equaled the product of 193,707,721 with 761,838,257,287. Hence Mersenne's conjecture was erroneous for the prime $p = 67$. But mathematicians are not easily put off. The story certainly does not end here.

We now know millions of larger primes, including a total of 39 Mersenne primes (as of June 2003). These vast discoveries are due to applying several new theoretical algorithms on powerful modern computers. Currently the largest known prime is the Mersenne prime, $2^{13,466,917} - 1$, a titanic prime of 4,053,946 decimal digits! It was discovered in 2001 through the collective effort of mathematicians and computer scientists who are part of GIMPS, the "great internet Mersenne prime search," combining the resources of tens of thousands of computers over the internet. (In fact, a total of about 13,000 computer years of computational time was required.) The central figures earning credit for this latest discovery are Michael Cameron, Scott Kurowski, and George Woltman. And of course they have simultaneously discovered an accompanying perfect number, the 8,107,892-digit number, $2^{13,466,916} \cdot (2^{13,466,917} - 1)$. Although it is widely expected, no one has been able to prove that there are infinitely many Mersenne primes. Surprisingly, it has not been shown even that $2^p - 1$ is composite for infinitely many primes p.

But the relationship between Mersenne primes and perfect numbers is even stronger than the result of Euclid's suggests. Euler proved a partial converse to Theorem 5.1 (published posthumously, 1848). It states

Theorem 5.2: *Let n be an even perfect number. Then n is of the form $2^{p-1}(2^p - 1)$ where $2^p - 1$ is a Mersenne prime.*

Proof of Theorem 5.2: *Let $n = 2^c \cdot b$ where $c \geq 1$ and b is odd. Since n is perfect, $\sigma(n) = 2n$. Since σ is a multiplicative function,*

$$\sigma(n) = \sigma(2^c)\sigma(b).$$

But

$$\sigma(n) = 2n = 2^{c+1} \cdot b.$$

Since $\sigma(2^c) = 2^{c+1} - 1$, we have

$$2^{c+1} \cdot b = (2^{c+1} - 1)\sigma(b).$$

Note that 2^{c+1} and $2^{c+1} - 1$ are relatively prime. Thus 2^{c+1} is a divisor of $\sigma(b)$. Hence we can write

$$\sigma(b) = d \cdot 2^{c+1}$$

for some positive integer d. Equivalently,

$$b = (2^{c+1} - 1)d.$$

If we can show that $d = 1$, then we have shown that n is of the form $2^{c+1} \cdot (2^{c+1} - 1)$.

Suppose, for the sake of contradiction, that $d > 1$. Then the number b has at least 1, b, and d as distinct divisors. It follows that

$$d \cdot 2^{c+1} = \sigma(b) \geq b + d + 1 = (2^{c+1} - 1)d + d + 1 = 2^{c+1} \cdot d + 1,$$

a clear contradiction.

All that remains is to demonstrate that $2^{c+1} - 1$ *is prime. But*

$$\sigma(2^{c+1} - 1) = \sigma(b) = 2^{c+1} = b + 1.$$

So $2^{c+1} - 1$ *is prime and hence* $c + 1 = p$*, a prime. Hence* $n = 2^{p-1}(2^p - 1)$ *where* $2^p - 1$ *is a Mersenne prime.* □

By combining Theorems 5.1 and 5.2, we have a complete characterization of even perfect numbers. But what about odd perfect numbers? No one has ever found such an example, nor has it been shown that none could exist. If there is such a beast, it must have at least eight prime factors (one of which is larger than ten million) and be at least 300 digits in length.

So what do you make of this example due to the great French philosopher and mathematician René Descartes (1596–1650)? Let $n = 3^2 \cdot 7^2 \cdot 11^2 \cdot 13^2 \cdot 22{,}021$. All factors listed are relatively prime. Hence

$$\sigma(n) = \sigma(3^2) \cdot \sigma(7^2) \cdot \sigma(11^2) \cdot \sigma(13^2) \cdot \sigma(22{,}021).$$

But $\sigma(3^2) = (3^3 - 1)/(3 - 1) = 13$, $\sigma(7^2) = (7^3 - 1)/(7 - 1) = 57$, $\sigma(11^2) = (11^3 - 1)/(11 - 1) = 133$, and $\sigma(13^2) = (13^3 - 1)/(13 - 1) = 183$. If 22,021 is prime, then $\sigma(22{,}021) = 22{,}021 + 1 = 22{,}022$. But

$$\begin{aligned} 13 \cdot 57 \cdot 133 \cdot 183 \cdot 22{,}022 &= 13 \cdot (3 \cdot 19) \cdot (7 \cdot 19) \cdot (3 \cdot 61) \cdot (2 \cdot 7 \cdot 11^2 \cdot 13) \\ &= 2 \cdot 3^2 \cdot 7^2 \cdot 11^2 \cdot 13^2 \cdot (19^2 \cdot 61) \\ &= 2 \cdot 3^2 \cdot 7^2 \cdot 11^2 \cdot 13^2 \cdot 22{,}021 = 2n. \end{aligned}$$

Is this an example of an odd perfect number? Of course not. Since the number $22{,}021 = 19^2 \cdot 61$, it is not prime and hence $\sigma(22{,}021) \neq 22{,}022$. So for $n = 3^2 \cdot 7^2 \cdot 11^2 \cdot 13^2 \cdot 22{,}021$, $\sigma(n) \neq 2n$. Even so, I trust you can appreciate the cleverness of Descartes's example.

The ancient Greeks were aware of an interesting pair of numbers, 220 and 284. The sum of the proper divisors of one gives the other, and vice versa. Such numbers are called *amicable pairs* (or friendly numbers). Since both Greek and Hebrew letters have numerical meaning as well, it was considered a propitious sign if your name or house number formed such a pair with a friend or associate. The search for further examples yielded no results until the Middle Ages. In the ninth century, the Arab scholar Thabit ibn-Qurra (826–901) discovered a remarkable formula to aid in the construction of amicable pairs. Here's what it states:

Theorem 5.3: *If* p*,* q*, and* r *are primes of the form* $p = 3 \cdot 2^{n-1} - 1$, $q = 3 \cdot 2^n - 1$*, and* $r = 9 \cdot 2^{2n-1} - 1$ *for some integer* $n > 1$*, then* $a = 2^n pq$ *and* $b = 2^n r$ *form an amicable pair.*

Amicable pairs of this form are now called Thabit pairs. Interestingly, Thabit ibn-Qurra himself did not find any Thabit pairs other than the well-known 220, 284 pair ($n = 2$). However, $n = 4$ yields the pair 17,296 and 18,416 as noted by the Moroccan al-Banna (1256–1321) and the Persian al-Farisi (c. 1260–1320) about the year 1300. (Fermat rediscovered this example in 1636.) The next example of a Thabit pair occurs for $n = 7$. Both Muhammed Baqir Yazdi (c.1600) and Descartes (1638) independently discovered it. But then the search for Thabit pairs stops since there are no other known examples where p, q, and r are simultaneously prime. Even so, there are plenty of amicable pairs of other forms. But first let's prove Thabit's result.

Proof of Theorem 5.3: *It suffices to show that both* $\sigma(a) = \sigma(b)$ *and that* $\sigma(b) = a + b$. *Now* $\sigma(a) = \sigma(2^n)\sigma(p)\sigma(q)$ *and* $\sigma(b) = \sigma(2^n)\sigma(r)$. *Hence it suffices to show that*

$$\sigma(p)\sigma(q) = \sigma(r) \text{and}$$

$$\sigma(2^n)\sigma(r) = 2^n(pq + r).$$

Since p, q, *and* r *are prime,*

$$\sigma(p)\sigma(q) = (p + 1)(q + 1) = 9 \cdot 2^{2n-1} = r + 1 = \sigma(r),$$

establishing the first equation.

To establish the second equation, note that

$$\begin{aligned}
\sigma(2^n)\sigma(r) &= (2^{n+1} - 1)(r + 1) \\
&= 2^n(9 \cdot 2^{2n} - 9 \cdot 2^{n-1}) \\
&= 2^n(9 \cdot 2^{2n} - 6 \cdot 2^{n-1} - 3 \cdot 2^{n-1}) \\
&= 2^n(9 \cdot 2^{2n-1} + 9 \cdot 2^{2n-1} - 3 \cdot 2^n - 3 \cdot 2^{n-1}) \\
&= 2^n(pq + r).
\end{aligned}$$

□

In 1750, Euler modified Theorem 5.3 from which he produced 62 more examples (plus, uncharacteristically, two erroneous examples). Euler's examples included the wonderful twin pairs 609,928, 686,072 and 643,336, 652,664, which share the same sum. Despite all these successes, it was a pleasant surprise when in 1866, a 16-year-old Italian student, B.N.I. Paganini, discovered the second smallest pair: 1,184 and 1,210.

Today, all amicable pairs with smaller number less than 10^{14} have been cataloged (David Einstein, 2003). There are precisely 39,374 such amicable pairs. In addition, over 2.6 million larger amicable pairs have been discovered. Each member of the largest known pair has over 5,000 digits. But are there infinitely many amicable pairs? Can amicable pairs be relatively prime to each other? Are there any amicable pairs of opposite parity (one odd and one even)? No one yet knows. Care to find out?

WORTH CONSIDERING

1. Verify that $2^{11} - 1$ is not prime by finding its two prime factors. Does this contradict Observation 1?
2. Use Fermat's result that factors of $2^p - 1$ are of the form $2np + 1$ to verify the compositeness of $2^{23} - 1$.
3. Show that $a^n - 1$ is composite if $a > 2$ and $n > 1$.
4. What is $\sigma(n)$ for $n = 100, 1{,}000, 22{,}021$?
5. Show that all even perfect numbers are triangular numbers.
6. **(a)** Verify that $28 = 1^3 + 3^3$ and that $496 = 1^3 + 3^3 + 5^3 + 7^3$.

   (b) Show that every even perfect number > 6 is the sum of consecutive odd cubes.
7. Find the single example of an odd abundant number less than 1,000.
8. Compute the three known Thabit pairs for $n = 2, 4$, and 7 in Theorem 5.3.
9. Find the mates for each of the following halves of amicable pairs: 2,620 (L. Euler, 1747), 6,232 (L. Euler, 1750), $122{,}368 = 2^9 \cdot 239$ (P. Poulet, 1941).
10. Let $s(n) = \sigma(n) - n$ denote the sum of the proper divisors of n. Define $s_k(n)$ recursively by $s_1(n) = s(n)$ and $s_k(n) = s(s_{k-1}(n))$ for $k > 1$. Call an integer n *sociable* of order k if $n = s_k(n)$ for some $k \geq 1$.

    (a) Verify that perfect numbers are sociable of order 1 and amicable pairs are sociable of order 2.

    (b) Verify that 12,496 is sociable of order 5 (P. Poulet). List four other numbers sociable of order 5.

    (Currently over 100 examples of sociable numbers of order 4 are known, but none of order 3.)
11. Prove the following two identities that appear in al-Banna's *Raf al-Hijab*:

    (a) $1^3 + 3^3 + \ldots + (2n-1)^3 = n^2(2n^2 - 1)$

    (b) $1^2 + 3^2 + \ldots + (2n-1)^2 = n(2n-1)(2n+1)/3$.

6 The Harmonic Series ... and Less

The motto of the ancient Pythagorean school was "all is number." To them the natural numbers and their ratios were the basis for all natural phenomena whether terrestrial or celestial. In music, the Pythagoreans studied the relationship between the lengths of stretched strings and the notes that they produced. For example, if the length of a string is halved, then the sound emitted by plucking the string goes up one octave. If a third of the string is pinched off, then the resulting note is a fifth above the octave. Such investigations led to the study of musical harmonics.

The *harmonic series* in the title of this chapter is the infinite sum $1 + \frac{1}{2} + \frac{1}{3} + \frac{1}{4} + \ldots$, which is denoted compactly by the mathematical notation $\sum_{n=1}^{\infty} \frac{1}{n}$. The notation indicates that we are to sum the quantity $1/n$ as n ranges from one to "infinity", that is, indefinitely. If we sum from $n = 1$ to $n = 5$, the sum is 2.28333 (to five decimal places) and if we sum from $n = 1$ to $n = 10$, the sum is 2.92897. Table 6.1 is a chart for the partial sum of the harmonic series for the first few powers of ten (each sum rounded to five significant digits). The sum seems to be growing rather slowly. For that reason it's natural to ask whether or not the sum levels off by *converging* to some limiting value. We will shortly return to this question after first providing some background discussion.

The Pythagoreans had a rival philosophical school, the Eleatics, who believed in the unity and permanence of nature and were bothered by the Pythagorean assumption that time and space were made up of multitudes of points and instants respectively. You no doubt have heard some of the paradoxes of Zeno (c. 450 B.C.E.), their most passionate proponent. Zeno's arguments were dialectical in that he began with his opponents assumptions and then through discussion, showed how they led to an absurdity. One famous example is the *Achilles*. Here it is argued that in order to traverse any given distance a runner must first traverse half of that distance. But before that he must traverse half of that, and so on. So he must complete an infinite number of steps in a finite amount of time. Surely this is impossible, or so it was argued, and hence the idea that space can be indefinitely subdivided is an absurdity. An alternative version of the Achilles story is that in order to traverse a given

Mathematical Journeys, by Peter D. Schumer
ISBN 0-471-22066-3

TABLE 6.1 Partial sums of the harmonic series

N	10	100	1,000	10,000	100,000	1,000,000
$\sum_{n \leq N} \frac{1}{n}$	2.92897	5.18738	7.48547	9.78761	12.09015	14.39273

distance, Achilles must traverse the first half of that distance. Then he must traverse half of the remaining distance. Then he must traverse the remaining half, ad infinitum. How could he ever finish the race in any finite amount of time?

A more modern way to view Achilles's race is to consider the total distance as comprising one unit. First he travels half the distance, then half of half (a quarter), then half of a half of a half (an eighth), etc. If we assume he runs at a constant pace, of course he finishes the race since it only takes him half as long to traverse each subsequent leg of his journey. In other words, the infinite sum $1+\frac{1}{2}+\frac{1}{4}+\frac{1}{8}+\ldots=\sum_{n=1}^{\infty}\frac{1}{2^n}$ converges to 1. This particular case is an example of what is known as a *geometric series*. So in some instances at least, an infinite addition can have a finite sum. An infinite series which steadily approaches a finite value is said to *converge* to that value. Other infinite series for which the partial sums increase without bound (or do not steadily approach a finite value) are said to *diverge*.

To aid in making the above discussion more concrete, recall that with a geometric series the ratio of successive terms is constant. So a geometric series is an infinite series of the form $a + ar + ar^2 + ar^3 + \ldots$ where $a \neq 0$ and r is a any constant. If $|r| < 1$, then the series converges (in fact to the quantity $\frac{a}{1-r}$). If $|r| \geq 1$, then the geometric series diverges. So the series $3+\frac{3}{4}+\frac{3}{16}+\ldots$ with $r=\frac{1}{4}$ converges to the value $\frac{3}{1-\frac{1}{4}}=4$. However, the series $1-2+4-8+16-\ldots$ with $r=-2$ diverges.

Zeno's arguments were subtle and clever and have led to a deeper understanding of the nature of time, space, and the infinite. But Zeno lacked a full understanding of the concept of convergence of infinite series. It was the study of these and similar ideas which led to the development of calculus, a process which took over 2,000 years.

Returning now to our harmonic series $\sum_{n=1}^{\infty}\frac{1}{n}$, is it a convergent series like the one in the Achilles story or does it diverge, that is, get larger and larger without bound. Despite our numerical evidence that suggests rather slow growth, it turns out that the harmonic series diverges. In fact, the Parisian scholar Nicole Oresme (1323–1382) convincingly demonstrated this as follows:

Proposition 6.1: *The harmonic series $\sum_{n=1}^{\infty}\frac{1}{n}$ diverges.*

Proof of Proposition 6.1: *We prove that the harmonic series diverges by showing that the sum is bounded below by a series that clearly diverges. To arrive at the smaller, comparison series we first associate (or congregate) 2^r successive terms of the harmonic series for each $r \geq 1$ beginning with the terms $\frac{1}{3}+\frac{1}{4}$. In* *particular,*

$$\sum_{n=1}^{\infty} \frac{1}{n} = 1 + \frac{1}{2} + \left(\frac{1}{3} + \frac{1}{4}\right) + \left(\frac{1}{5} + \frac{1}{6} + \frac{1}{7} + \frac{1}{8}\right) + \left(\frac{1}{9} + \ldots + \frac{1}{16}\right) + \left(\frac{1}{17} + \ldots + \frac{1}{32}\right) + \ldots$$

$$= 1 + \frac{1}{2} + \sum_{r=1}^{\infty} \left(\frac{1}{2^r + 1} + \ldots + \frac{1}{2^{r+1}}\right).$$

But $\frac{1}{2^r+1} + \ldots + \frac{1}{2^{r+1}} \geq \frac{1}{2^{r+1}} + \ldots + \frac{1}{2^{r+1}}$ *(consisting of* 2^r *identical terms) since each term of the sum on the left is at least as large as the corresponding term of the sum on the right.*

Furthermore, $\frac{1}{2^{r+1}} + \cdots + \frac{1}{2^{r+1}} = \frac{2^r}{2^{r+1}} = \frac{1}{2}$*. Thus,* $\sum_{n=1}^{\infty} \frac{1}{n} \geq 1 + \frac{1}{2} + \frac{1}{2} + \frac{1}{2} + \ldots$ *(with an infinite number of one-halves), which certainly increases without bound. Since the harmonic series is even larger, it must diverge as well.* □

There have been many other proofs that the harmonic series diverges. In a college calculus course, one is most likely to encounter a proof due to Jakob Bernoulli (1654–1705), which utilizes what is now known as the integral test to compare the harmonic series with that of the improper integral, $\int_1^{\infty} \frac{1}{x}\,dx$. The latter integral evaluates as the logarithm function, and the fact that the logarithm function grows without bound implies the divergence of the harmonic series. Here however is a simpler argument not requiring integral calculus.

Second Proof of Proposition 6.1: *Assume, contrary to what we wish to show, that the harmonic series converges. Say it converges to the real number s. Then the sum of the reciprocals of the even numbers,*

$$\sum_{n=1}^{\infty} \frac{1}{2n} = \frac{1}{2} \sum_{n=1}^{\infty} \frac{1}{n} = s/2.$$

Thus, the sum of the reciprocals of the odd numbers,

$$\sum_{n=1}^{\infty} \frac{1}{2n-1} = \sum_{n=1}^{\infty} \frac{1}{n} - \sum_{n=1}^{\infty} \frac{1}{2n} = s - s/2 = s/2 \text{ too.}$$

But $1 > \frac{1}{2}, \frac{1}{3} > \frac{1}{4}$*, and in general* $\frac{1}{2n-1} > \frac{1}{2n}$ *for all* $n \geq 1$*. So the notion that the sum of reciprocals of the even numbers has the same sum as the sum of reciprocals of the odd numbers is an absurdity. Therefore, the harmonic series cannot converge but rather must diverge.* □

Although the harmonic series diverges, the Bernoulli brothers, Jakob and Johann (1667–1749), had a thorough understanding of the convergence of other

p-series. The p-series are those of the form $\sum_{n=1}^{\infty} \frac{1}{n^p}$ for real $p > 0$. By the integral test, it was known that all p-series with $p > 1$ converge. But knowing that a series converges and determining what in fact it converges to are two entirely different matters. In fact, both Bernoullis tried in vain to determine the sum of the reciprocals of the squares

$$\sum_{n=1}^{\infty} \frac{1}{n^2} = 1 + \frac{1}{4} + \frac{1}{9} + \frac{1}{16} + \frac{1}{25} + \cdots.$$

Johann Bernoulli

Other noteworthy mathematicians worked on the problem as well. In England, John Wallis (1616–1703), the Savilian Professor of Geometry at Oxford University and one of the organizers of the Royal Society, helped further develop the theory of infinite series. In addition, he is credited with one of the most famous infinite products involving the number pi, namely

$$\frac{4}{\pi} = \frac{3}{2} \cdot \frac{3}{4} \cdot \frac{5}{4} \cdot \frac{5}{6} \cdot \frac{7}{6} \cdot \frac{7}{8} \cdot \frac{9}{8} \cdot \cdots.$$

Wallis also worked on determining the sum of the reciprocals of the squares, but could only make the numerical approximation

$$\sum_{n=1}^{\infty} \frac{1}{n^2} \approx 1.645.$$

Another mathematician, who had less talent but some impressive contacts, was Christian Goldbach (1690–1764). He was able to determine that

$$\frac{41{,}423}{25{,}200} < \sum_{n=1}^{\infty} \frac{1}{n^2} < \frac{76{,}997}{46{,}800}.$$

From this it follows that the series is bounded between 1.643769 and 1.645235, perhaps a slight improvement over Wallis's result.

Today, Goldbach is much more famous for his later number-theoretic conjecture, namely that every even number beyond 2 can be expressed as the sum of two prime numbers. In 1742, Goldbach reported his conjecture to the great Leonhard Euler. At first Euler felt that Goldbach's Conjecture must be either false or trivially true, but in time he came to realize that it was a very difficult problem indeed. In fact, Goldbach's Conjecture is an open problem to this very day! But now let's return to a decade earlier when Euler became intrigued with the sum of the reciprocals of the squares problem.

In 1731, Euler was able to equate the desired series with a sum involving a faster converging series. In this way, far fewer terms were required to evaluate the sum to any given degree of accuracy. In particular, Euler showed that

$$\sum_{n=1}^{\infty} \frac{1}{n^2} = (\log 2)^2 + \sum_{n=1}^{\infty} \frac{1}{2^{n-1} n^2}.$$

From this it readily followed that $\sum_{n=1}^{\infty} \frac{1}{n^2} \approx 1.644934$, accurate to six decimal places.

Euler, whom the contemporary French scholar François Arago called "Analysis Incarnate," continued to work on the problem and by 1733, through similar means and prodigious calculations, was able to derive the following incredible estimate:

$$\sum_{n=1}^{\infty} \frac{1}{n^2} \approx 1.64493406684822643647.$$

Even so, a closed form for the sum, if there even was one, seemed no closer now than when Jakob Bernoulli first considered the problem. But Euler was undeterred and in 1734 he made a spectacular breakthrough. What he discovered was that, in fact,

$$\sum_{n=1}^{\infty} \frac{1}{n^2} = \frac{\pi^2}{6}. \tag{6.1}$$

He was aware that his first proof of Formula 6.1 was not completely rigorous, but the result was certainly true as a quick calculation could verify. His reasoning was based on utilizing a theorem of Isaac Newton (1642–1727) concerning roots of polynomials and extending its application to infinite series (essentially polynomials of infinite degree). Let's take a brief aside to learn more.

Proposition 6.2 (Newton's Polynomial Roots Rule): *If $p(x)$ is a monic polynomial with constant term one, then the sum of the reciprocals of the roots of p is the negative of the coefficient of the linear term.*

Recall that a monic polynomial is one with leading coefficient of one and that the roots of p are the values of x for which $p(x) = 0$. For example, consider the polynomial $p(x) = x^3 + 3x^2 + 3x + 1 = (x + 1)^3$. The roots are -1, -1, and -1. The sum of the reciprocals of the roots is -3, the negative of the coefficient of x in $p(x)$ as expected by Newton's Rule.

Proof of Proposition 6.2: *Let $p(x) = x^n + a_{n-1}x^{n-1} + \ldots + a_1 x + 1 = (x - r_1)\cdots(x - r_n)$ where $r_1, \ldots, r_n$ are the not necessarily distinct roots of p guaranteed by the fundamental theorem of algebra.*

Multiplying the right-hand side and equating like coefficients we see that

$$1 = (-1)^n r_1 \cdots r_n \text{ and}$$

$$a_1 = \sum_{i=1}^{n} (-1)^{n-1} r_1 \cdots r_n / r_i.$$

Hence $a_1 = a_1/1 = -\sum_{i=1}^{n} \frac{1}{r_i}$. □

Euler knew that the function $y = \sin x$ can be represented by a power series that converges absolutely for all x (the Maclaurin series for $\sin x$). In fact, the well-known series is

$$\sin x = x - \frac{x^3}{3!} + \frac{x^5}{5!} - \frac{x^7}{7!} + \ldots = \sum_{n=1}^{\infty} (-1)^n \frac{x^{2n-1}}{(2n-1)!}. \tag{6.2}$$

Furthermore, the sine function has roots at every integral multiple of π, positive, negative, and zero. Suppose that r is a nonzero value for which $\sin r = 0$. It follows that

$$0 = r - \frac{r^3}{3!} + \frac{r^5}{5!} - \frac{r^7}{7!} + \ldots.$$

Since $r \neq 0$, we can divide both sides by $r : 0 = 1 - \frac{r^2}{3!} + \frac{r^4}{5!} - \frac{r^6}{7!} + \ldots$. Let $t = r^2 : 0 = 1 - \frac{t}{3!} + \frac{t^2}{5!} + \frac{t^3}{7!} + \ldots$. The coefficient of the linear term is $-1/6$. In addition, since the nonzero roots of $\sin x$ comprised all integral multiples of π, the roots of the series above are π^2, $(2\pi)^2$, $(3\pi)^2$, etc.

Now apply Newton's rule to the "infinite polynomial" above to obtain

$$1/6 = \frac{1}{\pi^2} + \frac{1}{(2\pi)^2} + \frac{1}{(3\pi)^2} + \dots.$$

By multiplying both sides by π^2, we get $\frac{\pi^2}{6} = \sum_{n=1}^{\infty} \frac{1}{n^2}$.

Euler had every reason to trust his keen insight, but he also knew that a rigorous proof could not involve applying a result to infinite series that should only apply to polynomials. Although we will not delve into the details here, Euler developed other infinite *product* representations for $\sin x$ and was able to prove the above result in a more rigorous manner. But he certainly didn't stop there. He defined the *zeta function* by $\zeta(s) = \sum_{n=1}^{\infty} \frac{1}{n^s}$ for $s > 1$. So the previous result could be restated as $\zeta(2) = \frac{\pi^2}{6}$.

Jakob Bernoulli had investigated formulas for the sums of squares, cubes, and so on. In doing so he introduced an infinite collection of numbers, now called the Bernoulli numbers B_n. They are defined by the following Taylor series

$$\frac{x}{e^x - 1} = \sum_{n=0}^{\infty} \frac{B_n}{n!} x^n \tag{6.3}$$

or alternatively by the following recurrence relation:

$$B_0 = 1 \text{ and } (n+1)B_n = -\sum_{k=0}^{n-1} \binom{n+1}{k} B_k \text{ for } n \geq 1. \tag{6.4}$$

Recall that the binomial coefficient $\binom{n+1}{k} = \frac{(n+1)!}{(n+1-k)!k!}$.

Euler derived a formula for the zeta function that applied to all positive even arguments, namely 2, 4, 6, and so on. Here is the result.

Euler's Theorem (1736): $\zeta(2k) = \frac{(-1)^{k+1}(2\pi)^{2k} B_{2k}}{2(2k)!}$. $\quad (6.5)$

Euler's beautiful result applies to any even value of the zeta function no matter how large. It follows that $\zeta(4) = \pi^4/90$ and $\zeta(6) = \pi^6/945$. Euler, who delighted in mental calculation, later tabulated many such values of the zeta function including the whopping

$$\zeta(26) = \frac{1315862}{11094481976030578125} \pi^{26}.$$

All of the zeta series that we have discussed can be thought of as a thinning out of our original harmonic series. Instead of getting a divergent series by adding up the reciprocals of all the natural numbers, we eliminate certain terms to see if a convergent series arises. So the question becomes how many terms and which ones need to be removed in order to obtain a convergent series. That's the theme of this chapter and the reason for the title, "The Harmonic Series ... and Less."

Certainly if we remove any finite number of terms from the harmonic series, we will still be left with a divergent series since an unbounded sum minus a finite number is still unbounded. In Chapter 1 we established that there are infinitely many primes. So what about the sum $\sum_p \frac{1}{p}$ where we sum over all primes p? It's an interesting question and one that Euler himself asked—and answered. Despite the fact we've eliminated all the composite numbers, the sum of the reciprocals of the primes diverges. In proving this result, he established another proof (independent of Euclid's) that there are infinitely many primes. We present a more elementary and modern proof of Euler's result.

Theorem 6.3 (Euler, 1737): *The series $\sum_p \frac{1}{p}$ diverges.*

Proof (James Clarkson, 1966): *Assume that $\sum_p \frac{1}{p}$ converges. Then there is an integer k such that $\sum_{m=k+1}^{\infty} \frac{1}{p_m} < \frac{1}{2}$ where p_m is the mth prime number. Now let $Q = p_1 \cdots p_k$ be the product of the first k primes. Consider the numbers $1 + nQ$ for $n = 1, 2, 3, \ldots$. None of these numbers is divisible by any of the first k primes (since each leaves a remainder of one upon division). Hence all the prime factors of $1 + nQ$ occur among the rest of the primes $p_{k+1}, p_{k+2}, \ldots$. Thus for all $r \geq 1$:*

$$\sum_{n=1}^{r} \frac{1}{1+nQ} \leq \sum_{t=1}^{\infty} \left(\sum_{m=k+1}^{\infty} \frac{1}{p_m} \right)^t$$

since the sum on the right includes among its terms all the terms on the left (and more). But $\sum_{m=k+1}^{\infty} \frac{1}{p_m} < \frac{1}{2}$. Thus

$$\sum_{t=1}^{\infty} \left(\sum_{m=k+1}^{\infty} \frac{1}{p_m} \right)^t < \sum_{t=1}^{\infty} \left(\frac{1}{2} \right)^t = 1.$$

(The last series is the geometric series we discussed in Zeno's Achilles Paradox). Hence the series $\sum_{n=1}^{\infty} \frac{1}{1+nQ}$ is dominated by the convergent geometric series $\sum_{t=1}^{\infty} \left(\frac{1}{2}\right)^t$, and hence $\sum_{n=1}^{\infty} \frac{1}{1+nQ}$ converges (by the Comparison Test). But, in fact, the series $\sum_{n=1}^{\infty} \frac{1}{1+nQ}$ diverges since $\sum_{n=1}^{\infty} \frac{1}{nQ}$ diverges (a constant times the harmonic series) and thus so does $\sum_{n=1}^{\infty} \frac{1}{1+nQ}$ by invoking the Limit Comparison Test. Therefore, the assumption that $\sum_p \frac{1}{p}$ converges is false. □

There are many ways to further thin out the primes and investigate the convergence of the sum of reciprocals of the remaining numbers. For example, in 1919, the Norwegian mathematician Viggo Brun developed a new and powerful "sieve method" to show that the sum of the reciprocals of the twin primes converges.

Recall that the twin primes are those that differ by two from another prime. So although no one to this day has been able to prove that there are infinitely many twin primes, they are qualitatively significantly less dense than the full set of primes.

Now let's thin out the harmonic series another way. Consider the *7-free numbers* consisting of all the natural numbers lacking the digit 7 in their base 10 representation: 1, 2, 3, ..., 6, 8, 9, ..., 16, 18, ..., 69, 80, 81, ... There appear to be plenty of them. So it is a natural question to enquire whether or not $\sum_{n\ 7\text{-free}} \frac{1}{n}$ converges. The following proposition may come as a mild surprise (known to A.J. Kempner, 1914, and may even predate him).

Proposition 6.4: $\sum_{n\ 7\text{-free}} \frac{1}{n}$ *converges.*

Proof: *Since all digits from 1 to 9 save for 7 itself is 7-free, there are 8 one-digit positive integers that are 7-free. There are $8 \cdot 9 = 72$ two-digit numbers that are 7-free. (The first digit can be any digit from 1 to 9 except 7 and the second digit can be any digit from 0 to 9 except 7.) Similarly, there are $8 \cdot 9 \cdot 9$ three-digit numbers that are 7-free, and in general, there are $8 \cdot 9^{n-1}$ 7-free n-digit numbers. Furthermore, if n is a one-digit number, then $1/n \leq 1$. If n is a two-digit number, then $1/n \leq 1/10$, and in general, if n is a k-digit number, then $1/n \leq 1/10^{k-1}$. Hence*

$$\sum_{n\ 7\text{-free}} \frac{1}{n} < (1 + \dots + 1) + \left(\frac{1}{10} + \dots + \frac{1}{10}\right) + \left(\frac{1}{100} + \dots + \frac{1}{100}\right) + \dots$$

$$= \sum_{k=1}^{\infty} 8 \cdot 9^{k-1} \cdot \left(\frac{1}{10}\right)^{k-1}$$

$$= \sum_{k=0}^{\infty} 8 \cdot \left(\frac{9}{10}\right)^{k-1}$$

The last expression is a convergent geometric series ($r = 9/10$) and hence $\sum_{n\ 7\text{-free}} \frac{1}{n}$ converges. □

Analogously, for any $r = 0, 1, \dots, 9$, the sum $S_r = \sum_{n\ r\text{-free}} \frac{1}{n}$ converges. In a 1979 article in the *American Mathematical Monthly*, R. Baillie, calculated each of these sums to 20 decimal places. Here we make a more modest tabulation (to three decimal places) (Table 6.2). It is worth considering why the sums get larger as r increases (think of 0 as being 10). Realize that the smaller the value of n, the larger the contribution of $1/n$.

It is an interesting corollary that the sum $S = \sum_{r=0}^{9} S_r$ must itself converge (even with all the repetition of terms). At first glance you might think that the above sum actually contains the harmonic series, but the harmonic series diverges, and hence we have a contradiction. But although the sum S contains the reciprocals of lots of natural numbers, it does not contain them all. It omits the set of

TABLE 6.2 Sum of reciprocals of r-free integers

r	1	2	3	4	5	6	7	8	9	0
S_r	16.177	19.257	20.570	21.327	21.835	22.206	22.493	22.726	22.921	23.103

all those integers whose expansion contain all ten-decimal digits. The first two such numbers are 1,023,456,789 and 1,023,456,798. This leads to the following corollary:

Corollary 6.4.1: *Let D = the set of all integers whose decimal representation contain all ten distinct digits. Then $\sum_{n \in D} \frac{1}{n}$ diverges.*

The thinning out of a convergent series must converge. So it follows that $\sum_{p\ 7\text{-free}} \frac{1}{p}$ must converge since we are merely summing over the 7-free primes. Similarly, the sum $\sum_{p\ 1\text{-free}} \frac{1}{p} + \sum_{p\ 2\text{-free}} \frac{1}{p} + \ldots + \sum_{p\ 0\text{-free}} \frac{1}{p}$ converges. But the sum of the reciprocals of all the primes diverges. Hence we note in closing, with D as defined in Corollary 6.4.1, the amazing result:

Corollary 6.4.2: $\sum_{p \in D} \frac{1}{p}$ *diverges.*

What makes Corollary 6.4.2 seem so counterintuitive is that the primes p in D seem so large (and hence $1/p$ so small). The first three primes in the set D are 10,123,457,689, 10,123,465,789, and 10,123,465,897. Yet the sum of the reciprocal of the squares $1 + 1/4 + 1/9 + 1/16 + \ldots$. converges! The explanation is that "most" primes are actually really big and contain all ten decimal digits. In fact, there are many more of them than there are squares (even though we somehow "know" and have no trouble naming lots of small squares). The problem is that we tend to live among the set of puny integers and generally ignore the vast infinitude of larger ones. How trite and limiting our view!

WORTH CONSIDERING

1. Show that if $|r| < 1$, then the geometric series $a + ar + ar^2 + \ldots$ converges to $\frac{a}{1-r}$.
2. Show that for $n \geq 2$, the sum $1 + \frac{1}{2} + \cdots + \frac{1}{n}$ is never an integer.
3. A celebrated puzzle problem involves a bug that flies nonstop back and forth between the front bumpers of two cars heading straight towards each other. If the bug flies at 120 miles per hour and starts off when the cars are two miles apart with each car traveling at 60 miles per hour, what is the total distance that the bug flies?
4. Given that $\sum_{n=1}^{\infty} \frac{1}{n^2} = \frac{\pi^2}{6}$, determine the sum $\sum_{n=1}^{\infty} \frac{1}{(2n-1)^2}$.

5. Use Equation 6.2 and the fact that $\sin \frac{\pi}{2} = 1$ to derive an interesting infinite series for the number 1.
6. Use Equation 6.4 to determine the first eight Bernoulli numbers.
7. If we define $B_n = \frac{N_n}{D_n}$ where N_n and D_n are relatively prime and $D_n > 0$, then a theorem due to Von Staudt and Clausen (1840) states that if n is even, D_n is the product of all the primes p where $(p - 1)|n$. Verify this result for D_2, D_4, D_6, and D_8.
8. Use Equation 6.5 to calculate $\zeta(k)$ for $k = 2$, 4, 6, and 8.
9. Use the fact that $\frac{1}{n(n+1)} = \frac{1}{n} - \frac{1}{n+1}$ to determine $\sum_{n=1}^{\infty} \frac{1}{t_n}$ where t_n is the nth triangular number.

7 Fermat Primes, the Chinese Remainder Theorem, and Lattice Points

In Chapter 1 we presented Euclid's classic proof that there are infinitely many primes. In Chapter 6, we stated that Euler had established that $\sum_p \frac{1}{p}$ diverges, thus providing an alternate proof that there are infinitely many primes. Just for the fun of it, we begin this chapter with two more proofs that there are infinitely many primes—one by Euler and a second by the Hungarian (later American) analyst and mathematical pedagogist, George Pólya (1887–1985).

Recall that a geometric series $\sum_{k=0}^{\infty} r^k$ converges if $|r| < 1$ and diverges otherwise. If it converges, then $\sum_{k=0}^{\infty} r^k = \frac{1}{1-r}$. Euler reasoned as follows:

Suppose that $p_1 = 2$, $p_2 = 3, \ldots, p_n$ are all the primes. Since $0 < \frac{1}{p_i} < 1$ for all i, the geometric series $\sum_{k=0}^{\infty} \frac{1}{p_i^k}$ converges, in fact to $\frac{1}{1-\frac{1}{p_i}}$. Hence

$$\prod_{i=1}^{n} \left(\sum_{k=0}^{\infty} \frac{1}{p_i^k} \right) = \prod_{i=1}^{n} \frac{1}{1 - \frac{1}{p_i}}. \tag{7.1}$$

The left-hand side of Equation 7.1 can be expanded to

$$\left(1 + \frac{1}{2} + \frac{1}{4} + \ldots\right)\left(1 + \frac{1}{3} + \frac{1}{9} + \ldots\right)\left(1 + \frac{1}{5} + \frac{1}{25} + \ldots\right)\cdots$$
$$\times \left(1 + \frac{1}{p_i} + \frac{1}{p_i^2} + \ldots\right) = 1 + \frac{1}{2} + \frac{1}{3} + \frac{1}{4} + \frac{1}{5} + \frac{1}{6} + \ldots = \sum_{m=1}^{\infty} \frac{1}{m}$$

by the Fundamental Theorem of Arithmetic (which states that every positive integer has a unique factorization as a product of primes). But this is the divergent harmonic series. Since the right-hand side of Equation 7.1 is presumed to be finite, we arrive at a contradiction. Therefore, there must be infinitely many primes.

Mathematical Journeys, by Peter D. Schumer
ISBN 0-471-22066-3

Euler proceeded to show that $\sum_p \frac{1}{p}$ diverges by using properties of the logarithm function, but the argument above already sufficed to establish the infinitude of primes.

Among Euler's mathematical motivations was the quest to derive many of the number-theoretic claims and conjectures made by the French jurist and "amateur" mathematician, Pierre de Fermat (1601–1665). Fermat had inscrutable intuition and Euler and others were gradually able to validate many of his mathematical pronouncements. However, in one famous example, at least, Euler ended up debunking Fermat's claim.

Fermat was interested in creating a formula that would generate only primes. It happens that all nonconstant polynomials $f(n)$ are composite for infinitely many values of n. So what about exponential functions? Let's consider the simplest candidate, $f(n) = 2^n + 1$. Fermat realized that if $n = ab$ with $a \geq 3$ odd, then $2^n + 1$ could be factored. In particular,

$$2^n + 1 = 2^{ab} + 1 = (2^a + 1)(2^{b(a-1)} - 2^{b(a-2)} + 2^{b(a-3)} - \ldots + 1),$$

and thus $2^n + 1$ is composite. Hence a better candidate for primality is $f(n) = 2^n + 1$ where n itself is only divisible by the prime 2. Fermat defined f_n as $2^{2^n} + 1$ and verified that $f_0 = 3$, $f_1 = 5$, $f_2 = 17$, $f_3 = 257$, and $f_4 = 65{,}537$ are all prime. Based on this evidence, Fermat conjectured that f_n is prime for all n.

Pierre de Fermat

This particular conjecture has not fared well. In 1732, Euler factored $f_5 = 4{,}294{,}967{,}297$ into $641 \cdot 6{,}700{,}417$ and then showed that each cofactor was prime. Euler first succeeded in proving that for $n > 2$, any factor of f_n was of the form $k \cdot 2^{n+2} + 1$. To factor f_5, he just needed to check primes of the form $128k + 1$. The first such prime is 257, which is not a factor of f_5. But the second prime of the form $128k + 1$ is 641, and it is a factor of f_5. Obviously more work was required to verify that the cofactor 6,700,417 is also prime.

Since then, much larger values of f_n have been studied. Realize that the notation f_n belies the rapid super exponential growth of the Fermat numbers. In 1880, after many months of labor, F. Landry (then aged 82) found that $f_6 = 274{,}177 \cdot 67{,}280{,}421{,}310{,}721$. It was another 90 years before the next Fermat number, f_7, was completely factored by Brillhart and Morrison with the aid of a modern computer. To date, no other Fermat number has been shown to be prime, quite contrary to Fermat's original conjecture. Currently f_n has been completely factored for $n \leq 11$ and it is known that f_n is composite for all $n \leq 32$. In addition, some sporadic factors are known for some truly huge Fermat numbers. For example, Göran Axelson (2002) has shown that $7{,}619 \cdot 2^{50{,}081} + 1$ divides $f_{50{,}078}$. Interestingly, no one has been able to prove either that there are infinitely many prime Fermat numbers or to prove that there are infinitely many composite Fermat numbers! Of course, at least one of these assertions must be true.

Fermat numbers do arise in other contexts, however. In 1796, the incomparable Carl Friedrich Gauss (1777–1855) was able to construct a regular 17-sided polygon using only straightedge and compass, a feat not accomplished by any of the great Greek geometers. Furthermore, Gauss proved that a regular polygon of N sides can be so constructed if and only if N is a power of two times a product of distinct Fermat primes. So, in theory at least, a regular 65,537-sided polygon can be constructed with basic Euclidean tools (something actually attempted over a ten-year period by a Professor Hermes and currently preserved in a trunk at the Mathematical Institute of Göttingen). Of course, such a multisided polygon would look *a lot* like a perfect circle.

Now we present Pólya's proof that there are infinitely many primes. To show that the primes are unbounded it suffices to find a nonterminating sequence $1 < a_0 < a_1 < a_2 < \ldots$ of pairwise relatively prime integers. As you may have guessed, one valid choice is to let $a_n = f_n$, the nth Fermat number. We establish that the Fermat numbers are pairwise relatively prime by establishing the following proposition:

Proposition 7.1: *The Fermat numbers satisfy the equation*

$$f_m = f_0 f_1 \cdots f_{m-1} + 2 \text{ for } m \geq 1. \tag{7.2}$$

Since f_n is odd for all n, if any prime $p | f_n$ for some $n \leq m - 1$, then $p \nmid f_m$ by Proposition 7.1 (since $p \nmid 2$), and hence the Fermat numbers are pairwise relatively prime.

Proof of Proposition 7.1: *(Induction on m) Let $m = 1$. Then $f_1 - 2 = 2^{2^1} + 1 - 2 = 5 - 2 = 3 = f_0$.*

Now assume that the proposition holds for m, that is, that $f_m = f_0 f_1 \cdots f_{m-1} + 2$. Next we establish Equation 7.2 for the case $m + 1$.

$$\begin{aligned} f_{m+1} &= 2^{2^{m+1}} + 1 = (2^{2^m})^2 + 1 \\ &= [(2^{2^m} + 1)^2 - 2 \cdot 2^{2m} - 1] + 1 \\ &= (2^{2^m} + 1)^2 - 2(2^{2^m} + 1) + 2 \\ &= f_m^2 - 2f_m + 2 \\ &= f_m(f_m - 2) + 2 \\ &= (f_0 \cdots f_{m-1})f_m + 2 \text{ by the inductive hypothesis.} \end{aligned}$$

□

Next we turn to the Chinese Remainder Theorem, variants of which were known in ancient times. For example, the Chinese mathematician Sun-Tzi (ca. 300 C.E.) proposed the following problem: "There are things of an unknown number which when divided by 3 leave 2, by 5 leave 3, and by 7 leave 2. What is the number?" In general, there will be a solution as long as the divisors in the problem (in this example, 3, 5, and 7) are pairwise relatively prime. We state the theorem below:

Chinese Remainder Theorem: *Assume that $m_1, m_2, \ldots, m_r$ are pairwise relatively prime. Let $b_1, b_2, \ldots, b_r$ be any integers. Then the system of congruences*

$$\begin{cases} x \equiv b_1 \pmod{m_1} \\ x \equiv b_2 \pmod{m_2} \\ \quad\vdots \\ x \equiv b_r \pmod{m_r} \end{cases}$$

has a simultaneous solution for x. In fact, it is unique modulo $m_1 m_2 \cdots m_r$.

The Chinese Remainder Theorem is the basis for some arithmetic feats. For example, you could ask someone to pick a number from 1 to 1,000 (or even 1,001), then ask three other volunteers to tell the remainder of the chosen number upon division by 7, 11, and 13 in turn. Then, after a short calculation, you can correctly announce the original number. Here's what's involved:

Notice that

$$\begin{aligned} -77 &\equiv 0 \pmod{7}, \equiv 0 \pmod{11}, \text{ and } \equiv 1 \pmod{13} \\ 364 &\equiv 0 \pmod{7}, \equiv 1 \pmod{11} \text{ and } \equiv 0 \pmod{13} \\ -286 &\equiv 1 \pmod{7}, \equiv 0 \pmod{11}, \text{ and } \equiv 0 \pmod{13}. \end{aligned} \tag{7.3}$$

These three numbers will serve as our multipliers once our volunteers announce their results.

For instance, suppose that the number N is chosen and it is reported that N has remainder 3 when divided by 7, remainder 4 when divided by 11, and remainder 10 when divided by 13. Since 7, 11, and 13 are pairwise relatively prime, the Chinese Remainder Theorem guarantees a unique solution modulo $7 \cdot 11 \cdot 13 = 1{,}001$. In addition, in order to work with the smallest numbers possible (in absolute value), note that $10 \equiv -3 \pmod{13}$. By the way we've chosen our multipliers, we simply multiply the given remainders by the appropriate multipliers and sum. In this case we get

$$N \equiv 3(-286) + 4(364) - 3(-77) (\text{mod } 1{,}001). \text{ Hence}$$

$$N \equiv 829 \ (\text{mod } 1{,}001).$$

But $1 \leq 829 \leq 1{,}000$ and so N must be 829.

Recall Theorem 1.2 from Chapter 1, namely that there are arbitrarily large gaps between successive primes. Equivalently, for any natural number r there is a string of r consecutive composite numbers. In fact, the string $(r+1)!+2$, $(r+1)!+3, \ldots, (r+1)!+(r+1)$ is such an example.

Now we use the Chinese Remainder Theorem to demonstrate that there exist arbitrarily long strings of *highly* composite integers. By *highly composite* we mean having as many prime factors as we desire. Here is the result:

Theorem 7.2: *Let r and n be positive integers. There exist r consecutive numbers each divisible by at least n distinct primes.*

Proof of Theorem 7.2: *Let*

$$\begin{cases} p_1, \ldots, p_n \text{ be the first } n \text{ primes} \\ p_{n+1}, \ldots, p_{2n} \text{ be the second } n \text{ primes} \\ \qquad \vdots \\ p_{(r-1)n+1}, \ldots, p_{rn} \text{ be the } r^{\text{th}} \text{ set of } n \text{ primes} \end{cases}$$

Let

$$\begin{cases} m_1 = p_1 \cdots p_n \\ m_2 = p_{n+1} \cdots p_{2n} \\ \qquad \vdots \\ m_r = p_{(r-1)n+1}, \ldots, p_{rn} \end{cases}$$

Consider the simultaneous congruences

$$\begin{cases} x \equiv -1 \pmod{m_1} \\ x \equiv -2 \pmod{m_2} \\ \qquad \vdots \\ x \equiv -r \pmod{m_r} \end{cases}$$

The Chinese Remainder Theorem guarantees a solution N (mod $m_1 \cdots m_r$). Hence $m_1|(N+1), m_2|(N+2), \ldots,$ and $m_r|(N+r)$. So the r numbers, $N+1, N+2, \ldots, N+r$, are each divisible by at least n primes. □

Our last topic in this chapter deals with lattice points in the plane. *Lattice points* are those points in the plane having both x and y Cartesian coordinates equal to integers. If we look out from the origin and could see indefinitely far, then we would see infinitely many lattice points, namely all those points (a, b) where a and b are relatively prime. However, all the points (a, b) where a and b are not relatively prime would not be visible. This is readily apparent (mentally, if not visibly). Suppose that $\gcd(a, b) = d > 1$. Then the visible point $(\frac{a}{d}, \frac{b}{d})$ would block (a, b) from our view. Figure 7.1 shows a small part of the plane with light points being visible lattice points and dark points representing those that are not visible from the origin.

One interesting mathematical result related to Figure 7.1 is due to Gauss. For any real number $r > 0$, consider the square consisting of that portion of the Cartesian plane for which $|x| < r$ and $|y| < r$. Let $N(r)$ denote the total *number* of lattice points inside the square and let $V(r)$ denote the number of *visible* lattice points inside the square. Let $[r]$ denote the greatest integer less than or equal to r. On the one hand, it is easy to see that $N(r) = (2[r]+1)^2$ since there is a lattice point for all integers x and y with $-[r] \leq x \leq [r], -[r] \leq y \leq [r]$. On the other hand, there is no simple formula for $V(r)$. However, Gauss showed that $\lim_{r\to\infty} \frac{V(r)}{N(r)} = \frac{6}{\pi^2}$. (It is no coincidence that this number is the reciprocal of $\zeta(2)$ which we discussed in Chapter 6.) Since visible lattice points correspond to those points having relatively prime x and y values, we can roughly think

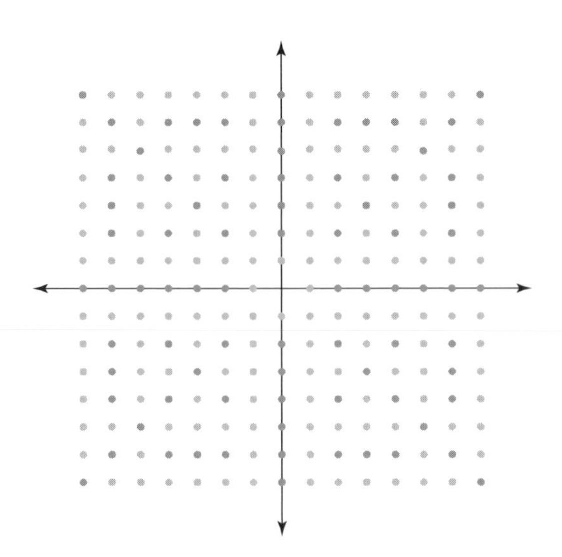

Figure 7.1 Green lattice points are visible from the origin, red lattice points are invisible.

of Gauss's result in the following way. The chance that two randomly chosen integers are relatively prime is $\frac{6}{\pi^2} \approx 0.6079$.

We now state another interesting theorem concerning visible lattice points.

Theorem 7.3: *The set of lattice points in the plane visible from the origin contains arbitrarily large square gaps. In particular, given any integer $n > 0$, there is a lattice point (a, b) for which all of the lattice points $(a+r, b+s)$, $0 < r \leq n$, $0 < s \leq n$ are invisible from the origin.*

Proof of Theorem 7.3: *As in the proof of Theorem 7.2, list the first n^2 primes in an $n \times n$ tableau:*

$$\begin{pmatrix} p_1 & p_2 & \cdots & p_n \\ p_{n+1} & p_{n+2} & \cdots & p_{2n} \\ \cdots & \cdots & \cdots & \cdots \\ p_{n(n-1)+1} & p_{n(n-1)+2} & \cdots & p_{n^2} \end{pmatrix}$$

Figure 7.2 Tableau of first n^2 primes.

Multiplying horizontally, let $m_r = p_{(r-1)n+1}p_{(r-1)n+2}\cdots p_{rn}$ for $1 \leq r \leq n$.
Multiplying vertically, let $M_s = p_s p_{n+s}\cdots p_{n(n-1)+s}$ for $1 \leq s \leq n$.
Notice that the m_r's are pairwise relatively prime, as are the M_s's.
Now consider the following congruence system:

$$\begin{cases} x \equiv -1 \pmod{m_1} \\ x \equiv -2 \pmod{m_2} \\ \qquad\vdots \\ x \equiv -n \pmod{m_n} \end{cases}$$

By the Chinese Remainder Theorem, there is a unique solution, $x \equiv a \pmod{m_1 m_2 \cdots m_n}$.

Similarly, the congruence system

$$\begin{cases} y \equiv -1 \pmod{M_1} \\ y \equiv -2 \pmod{M_2} \\ \qquad\vdots \\ y \equiv -n \pmod{M_n} \end{cases}$$

has a unique solution $y \equiv b \pmod{M_1 M_2 \cdots M_n}$. But $m_1 m_2 \cdots m_n = M_1 M_2 \cdots M_n$.

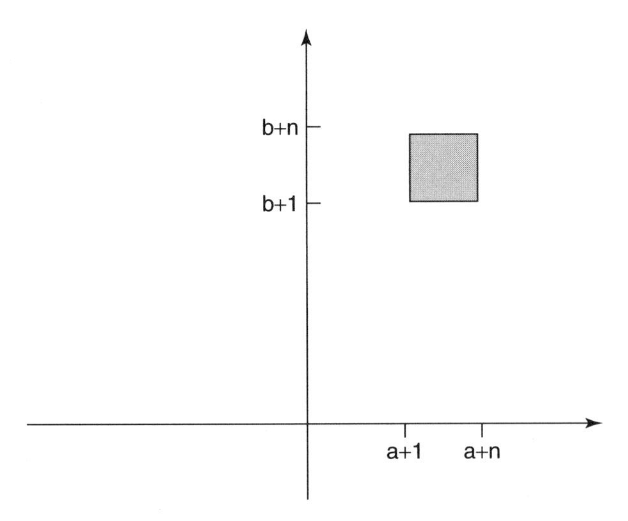

Figure 7.3 An invisible square.

Now consider the square with opposite vertices at $(a+1, b+1)$ *and* $(a+n, b+n)$ *(Fig. 7.3).*

Any lattice point inside or on the square has coordinates of the form $(a+r, b+s)$ *where* $0 < r \leq n$ *and* $0 < s \leq n$. *But by our construction,* $a \equiv -r \pmod{m_r}$ *and* $b \equiv -s \pmod{M_s}$. *Thus* $m_r|(a+r)$ *and* $M_s|(b+s)$. *But the prime* p *at the intersection of row* r *and column* s *in Figure 7.2 (namely* $p = p_{(r-1)n+s}$*) divides both* m_r *and* M_s. *So* $p|(a+r)$ *and* $p|(b+s)$. *Hence* $gcd(a+r, b+s) > 1$. *Since* r *and* s *were arbitrary, no point inside or on the square is visible.* □

Example: Let $M = \begin{bmatrix} 2 & 3 & 5 \\ 7 & 11 & 13 \\ 17 & 19 & 23 \end{bmatrix}$.

We readily calculate $m_1 = 30$, $m_2 = 1{,}001$, $m_3 = 7{,}429$, $M_1 = 238$, $M_2 = 627$, and $M_3 = 1{,}495$.

In this case, $m_1 m_2 m_3 = M_1 M_2 M_3 = 223{,}092{,}870$. Now let

$$\begin{cases} a \equiv -1 \pmod{m_1} \\ a \equiv -2 \pmod{m_2} \\ a \equiv -3 \pmod{m_3} \end{cases}$$

We get $a = 11{,}9740{,}619$.

Next let

$$\begin{cases} b \equiv -1 \pmod{M_1} \\ b \equiv -2 \pmod{M_2} \\ b \equiv -3 \pmod{M_3} \end{cases}$$

We obtain $b = 12{,}1379{,}047$.

So the 3×3 square with opposite vertices at $(a+1, b+1)$ and $(a+3, b+3)$ is an invisible 3×3 square.

Of course, there are invisible 3×3 squares closer to the origin, but the method we have presented generalizes for squares of any size. For the record, though, St. Michael's College computer scientist John Trono has found an invisible 3×3 square with southwest corner at (1,274, 1,308). To extend our method to find an invisible 4×4 square would involve calculations modulo $2 \cdot 3 \cdot 5 \cdot 7 \cdot 11 \cdot 13 \cdot 17 \cdot 19 \cdot 23 \cdot 29 \cdot 31 \cdot 37 \cdot 41 \cdot 43 \cdot 47 \cdot 53 = 32{,}589{,}158{,}477{,}190{,}044{,}730$. Care to find a closer 4×4 square?

WORTH CONSIDERING

1. Show that there are infinitely many primes of the form $4k+3$ by considering the number $N = 4p_1p_2 \cdots p_r - 1$ where the p_i's are distinct primes of that form.

2. J.F.T. Pepin (1826–1904) proved that the Fermat number f_n is prime if and only if $3^{(f_n-1)/2} \equiv -1 \pmod{f_n}$. Use Pepin's test to verify that f_2, f_3, and f_4 are prime.

3. Solve the remainder problem posited by Sun-Tzi.

4. **(a)** Find integers m_1, m_2, and m_3 for which

$$m_1 \equiv 1 \pmod 3, 0 \pmod 5, 0 \pmod 7$$

$$m_2 \equiv 0 \pmod 3, 1 \pmod 5, 0 \pmod 7$$

$$m_3 \equiv 0 \pmod 3, 0 \pmod 5, 1 \pmod 7.$$

(b) Use part (a) to find an integer between 1 and 100 inclusive that has remainder 2 when divided by 3, remainder 1 when divided by 5, and is divisible by 7.

(c) Find the smallest integer greater than 1,000 that has remainder 2 when divided by 3, remainder 3 when divided by 5, and remainder 4 when divided by 7.

5. Use Formula 7.3 to find the smallest positive integer that is $\equiv 1 \pmod 7$, $0 \pmod{11}$, and $5 \pmod{13}$?

6. What is the closest 2×2 square that is invisible from the origin.

7. Verify that the 3×3 square with southwest corner (1,274, 1,308) is invisible from the origin.

8 Tic-Tac-Toe, Magic Squares, and Latin Squares

Mathematical games and amusements can be intellectually stimulating and fun. As you get older and/or more sophisticated some of the easier games lose their interest. One example of this is tic-tac-toe, where two competitors alternately place X's and O's in a 3×3 matrix until one of them completes three in a row, vertically, horizontally, or diagonally. After a fair number of games, it becomes apparent that perfect play by both players always results in a draw. Even so, the strategy of tic-tac-toe can be applied to seemingly more intricate games as we will see shortly.

A more interesting game is the Fifteen Game. Nine cards are removed from a deck of playing cards—the ace through nine of some suit. The cards are laid out face up, available to each of two players. The players take turns choosing a card from the central pile and adding it to their own collection. The first player to have three cards that total 15 wins. Try a few rounds of this game with a friend and see how you do. Can you come up with any good strategies?

Here's another good game. I call it Eat Bee, but it's actually a version of the game Hot, invented by the Canadian mathematician, Leo Moser. As in the Fifteen Game, nine cards are laid out in a central pile and two players alternate in selecting cards. This time the cards have words written on them—Eat, Bee, Less, Via, Bits, Lily, Soda, Boo, and Loot. The first player to obtain three cards sharing a common letter wins the game. Give it a try.

Now we turn briefly to three by three magic squares. It may seem that we're jumping around haphazardly, but soon we'll see how everything relates together. A *magic square* is an $n \times n$ array of the numbers $1, 2, \ldots, n^2$ where each column and row adds up to the same sum (called the magic sum). Magic squares have been known for millennia and at one time may have been thought to have magic powers. For example, the magic square (Fig. 8.1) was known to the Emperor Yu of China who reigned approximately 2200 B.C.E. and purportedly saw it written on the back of a divine turtle. Notice that in this case, each row and column (and even the two main diagonals) add up to 15. By the way, it's very easy to remember this magic square if you just recall the placement of the numbers 1,

Mathematical Journeys, by Peter D. Schumer
ISBN 0-471-22066-3

$$\begin{bmatrix} 4 & 9 & 2 \\ 3 & 5 & 7 \\ 8 & 1 & 6 \end{bmatrix}$$

Figure 8.1 Oldest known magic square called the Lu Shu.

4, 9, 2 (think Christopher Columbus). Once those four numbers are placed, the rest of the numbers are forced into their proper slots.

We now see the best strategy for the Fifteen Game. The magic square in Figure 8.1 includes the nine card values in the Fifteen Game. The eight different winning hands in the Fifteen Game are listed as its rows, columns, and main diagonals. Hence playing the Fifteen Game is identical to playing tic-tac-toe on this magic square. Since perfect play in tic-tac-toe results in a draw, the same is true of the Fifteen Game. In practice, of course, your opponent might not know all of this and you should still have an advantage.

What about the Eat Bee game? It's the same story all over again. List the words as shown in Figure 8.2 and then play tic-tac-toe on the array of words. Note that there is a row for E, a row for I, and one for O. There's columns for A, B, and L and finally diagonals for S and T. Again the game should end in a draw.

Before moving on, take another look at Figure 8.1. If we read the numbers left to right as if they were three-digit numbers, then the three rows become 492, 357, and 816. If we read the numbers backwards we obtain 294, 753, and 618. It turns out that the sum of the squares of the two sets of numbers are equal; namely that $492^2+357^2+816^2 = 294^2+753^2+618^2$. In addition, we can do the same for the three columns. That is, $438^2+951^2+276^2 = 834^2+159^2+672^2$. Maybe that turtle was divine after all!

We now discuss magic squares in a bit more detail. An $n \times n$ magic square contains the numbers 1 through n^2 arranged in n rows. Since the sum $1+2+\ldots+n^2 = \frac{n^2(n^2+1)}{2}$ and each row has the same sum, the magic sum of an $n \times n$ magic square must be $\frac{n(n^2+1)}{2}$. For example, a 3×3 magic square has magic sum 15. Similarly, a 4×4 magic square has magic sum 34. An interesting example of such a square appears in the work of the German painter and engraver Albrecht Dürer (1471–1528). In his work, "Melancholia," along with a host of mathematical and scientific tools, the magic square shown in Figure 8.3 appears. Note that the date of the engraving appears in the middle of the bottom row! And this square has some extra magic as noted by the mathematician, Hossein Behforooz. If we add

$$\begin{bmatrix} \textit{EAT} & \textit{BEE} & \textit{LESS} \\ \textit{VIA} & \textit{BITS} & \textit{LILY} \\ \textit{SODA} & \textit{BOO} & \textit{LOOT} \end{bmatrix}$$

Figure 8.2 Strategy square for Eat Bee game.

$$\begin{bmatrix} 16 & 3 & 2 & 13 \\ 5 & 10 & 11 & 8 \\ 9 & 6 & 7 & 12 \\ 4 & 15 & 14 & 1 \end{bmatrix}.$$

Figure 8.3

the numbers along the two main diagonals, the sum is easily seen to be the same as the sum of the rest of the entries. So $16+10+7+1+13+11+6+4=68=3+2+8+12+14+15+9+5$. However, it is also true that the sum of the squares of the terms have equal sums; that is, $16^2+10^2+7^2+1^2+13^2+11^2+6^2+4^2=748=3^2+2^2+8^2+12^2+14^2+15^2+9^2+5^2$. And, believe it or not, the same is true for the sum of cubes: $16^3+10^3+7^3+1^3+13^3+11^3+6^3+4^3=9,248=3^3+2^3+8^3+12^3+14^3+15^3+9^3+5^3$. Dürer's artistry seems to extend well beyond his woodblocks.

One might think that the larger the square, the more difficult it is to construct such an example. But actually the number of distinct magic squares grows very rapidly as the size of the square increases. In fact, there are 880 different 4×4 magic squares, as was shown by the French mathematician Frenicle de Bessy (1605–1675). We will discuss the creation of another such square when we discuss Latin squares.

Magic squares are often classified as being either odd-ordered or even-ordered depending on whether the number of rows is odd or even respectively. There is a very convenient algorithm for creating odd-ordered magic square which we now describe. It is due to De la Loubère from the late 17th century. Here is the algorithm for creating an *odd-ordered* magic square of any given size:

1. Write the number 1 in the center square of the top row.
2. Each successive number is written by moving one space in a northeast direction (up and to the right) to the next vacant square. But you must think of the magic square as being written on a torus (the surface of a doughnut). Hence if you move off the magic square to the right, you show up in the same row on the left. If you move off the top of the magic square, you arrive in the same column at the bottom.
3. If you are at the top right corner or if the next square is already full, then move down one space from your last entry. Again, if you're already in the bottom row, then you must move "down" to the same column in the top row.

Example (3 x 3 magic square): *Using De la Loubère's algorithm we obtain the following square shown in figure 8.4, equivalent to our first example.*

Example (5 x 5 magic square): *Figure 8.5 gives an example of a 5×5 magic square.*

$$\begin{bmatrix} 8 & 1 & 6 \\ 3 & 5 & 7 \\ 4 & 9 & 2 \end{bmatrix}$$

Figure 8.4 3 × 3 magic square.

$$\begin{bmatrix} 17 & 24 & 1 & 8 & 15 \\ 23 & 5 & 7 & 14 & 16 \\ 4 & 6 & 13 & 20 & 22 \\ 10 & 12 & 19 & 21 & 3 \\ 11 & 18 & 25 & 2 & 9 \end{bmatrix}$$

Figure 8.5 5 × 5 magic square.

To be honest, magic squares are generally considered rather uninteresting by most serious mathematicians. After all they don't seem to be especially applicable to anything else. However, in the summer of 1988, for the start of a sabbatical my father and I drove cross-country from Middlebury, Vermont, to Sunnyvale, California. At a diner in Joliet, Illinois, our waitress noticed me doodling some mathematics on a napkin. She came over and said, "If you want a real tough problem, write out the numbers from 1 to 9 in a square so that all the rows and columns and diagonals add up to same thing. I'll give you until I come back for your order." Of course, we just had to write out a 3×3 magic square as in Figure 8.4. She was so impressed with how quickly we solved the problem, we each had apple pie à la mode for dessert—on the house. Now that's an application of magic squares, isn't it?

Benjamin Franklin

I now mention two of my favorite examples of magic squares. The first one is due to the celebrated American scientist and statesman Benjamin Franklin (1706–1790). As reported by him, he spent a fair amount of time as a young clerk for the Pennsylvania Assembly amusing himself by making such squares. Figure 8.6 shows one of his examples, and I think you'll agree he had quite a talent for it.

There are many interesting patterns within Franklin's 8×8 magic square. In addition to all rows and columns adding up to 260, if we break up the full square into four 4×4 quarter squares, then each of the quarter squares are pseudomagical in the sense that each has equal row and column sums of 130. The sum of the main diagonal going from the northwest corner to the southeast corner is 252. The sum of the other main diagonal going from the northeast corner to the southwest corner is 268. Interestingly, the sum of all the "broken" diagonals starting from the top (or bottom or left or right) and moving diagonally up or down to the left or to the right is also always 252 or 268 in an alternating pattern. A broken diagonal must still include one entry from each row and column. Again think of the square as being drawn on a torus. For example, $47 + 19 + 20 + 44 + 11 + 55 + 56 + 16 = 268 = 8 + 55 + 59 + 12 + 36 + 19 + 31 + 48$. Analogously, $21 + 38 + 42 + 25 + 49 + 2 + 14 + 61 = 252 = 33 + 18 + 62 + 13 + 5 + 54 + 26 + 41$.

Figure 8.7 shows my other favorite. This is an order eight magic square due to Leonhard Euler. What makes Euler's magic square so endearing is one special property it has in addition to possessing equal row and column sums. The numbers 1, 2, 3, ..., 64 form a knight's tour of the chessboard! Each successive number is an "L" step from the preceding number. Knight's tours had been studied by mathematicians such as Brook Taylor (1685–1731) and Abraham DeMoivre (1667–1754), but in 1759 Euler vastly extended their work and determined for which values of n such tours existed on $n \times n$ chessboards. Even so, the combination of magic square and knight's tour displayed in Figure 8.7 is quite striking.

$$\begin{bmatrix} 17 & 47 & 30 & 36 & 21 & 43 & 26 & 40 \\ 32 & 34 & 19 & 45 & 28 & 38 & 23 & 41 \\ 33 & 31 & 46 & 20 & 37 & 27 & 42 & 24 \\ 48 & 18 & 35 & 29 & 44 & 22 & 39 & 25 \\ 49 & 15 & 62 & 4 & 53 & 11 & 58 & 8 \\ 64 & 2 & 51 & 13 & 60 & 6 & 55 & 9 \\ 1 & 63 & 14 & 52 & 5 & 59 & 10 & 56 \\ 16 & 50 & 3 & 61 & 12 & 54 & 7 & 57 \end{bmatrix}.$$

Figure 8.6 Franklin Square (not a street address).

$$\begin{bmatrix} 1 & 48 & 31 & 50 & 33 & 16 & 63 & 18 \\ 30 & 51 & 46 & 3 & 62 & 19 & 14 & 35 \\ 47 & 2 & 49 & 32 & 15 & 34 & 17 & 64 \\ 52 & 29 & 4 & 45 & 20 & 61 & 36 & 13 \\ 5 & 44 & 25 & 56 & 9 & 40 & 21 & 60 \\ 28 & 53 & 8 & 41 & 24 & 57 & 12 & 37 \\ 43 & 6 & 55 & 26 & 39 & 10 & 59 & 22 \\ 54 & 27 & 42 & 7 & 58 & 23 & 38 & 11 \end{bmatrix}$$

Figure 8.7 Euler's Knight's tour magic square.

In a similar vein, mathematicians, computer scientists, and various puzzlists have sought to exhibit a magic square *where the diagonal sums are also magic* that forms a knight's tour of an 8x8 chessboard. Recently software written by J.C. Meyrignac was widely distributed via a website organized by Guenter Stertenbrink. A total of 140 distinct magic knight's tours with magic row and column sum were eventually found. However, after the equivalent of over 138 days of computation at a rate of 1 gigahertz (GHz), in August of 2003 it was determined that no magic knight's tour with the same row, column, and diagonal sums is possible. (The extremely fatigued horse is happy to hear about this and is now comfortably resting in his stable.)

Another interest of Euler's was the study of what he called Latin squares. A *Latin square* of order n is an $n \times n$ matrix where each row and column contains every natural number from 1 to n. Euler originally used Latin letters rather than numerals and hence the Latin square moniker has remained. Figure 8.8 contains two examples.

It is quite easy to construct a Latin square, but the example of the pair of Latin squares in Figure 8.8 has an additional property. They form a pair of orthogonal Latin squares. In general, denote the entries of square A by (a_{ij}) and the entries of square B by (b_{ij}) where i represents the row of the square and j its column. A and B are *orthogonal Latin squares* of order n if for every ordered pair (s, t) with $1 \leq s \leq n, 1 \leq t \leq n$, there is a location ij (i^{th} row, j^{th} column) such that $(a_{ij}, b_{ij}) = (s, t)$. That is, all n^2 possible ordered pairs actually occur.

$$A = \begin{bmatrix} 1 & 2 & 3 & 4 \\ 2 & 1 & 4 & 3 \\ 3 & 4 & 1 & 2 \\ 4 & 3 & 2 & 1 \end{bmatrix} \qquad B = \begin{bmatrix} 1 & 2 & 3 & 4 \\ 3 & 4 & 1 & 2 \\ 4 & 3 & 2 & 1 \\ 2 & 1 & 4 & 3 \end{bmatrix}$$

Figure 8.8 A pair of orthogonal Latin squares.

Check it on the example in Figure 8.8. By the way, the superposition of a pair of orthogonal Latin squares is sometimes called a Graeco-Latin square since Euler used Latin letters for one square and Greek letters for the other.

Latin squares have proven to have real-world applications. In agriculture, different fertilizers can be tested on different sets of soil in a manner designed by an appropriate-sized Latin square to see which combination works best in growing, say, a particular variety of corn. If we want to test several varieties of corn with the different fertilizers, then a Graeco-Latin square provides the experimental design. Similarly, Latin squares have found use in medical experiments and drug testing. In the world of pure mathematics, Latin squares are an integral part of the study of finite projective planes.

Here's how Euler utilized Latin squares to create magic squares. From Figure 8.8, convert the ordered pairs into two-digit numbers (forming the Graeco-Latin square below):

$$\begin{bmatrix} 11 & 22 & 33 & 44 \\ 23 & 14 & 41 & 32 \\ 34 & 43 & 12 & 21 \\ 42 & 31 & 24 & 13 \end{bmatrix}$$

Notice that by our construction, each number is unique and each row and column has the same sum, in this case 110. This is a pseudomagic square since the entries are not the first 16 numbers.

However, it is now an easy matter to transform it into a bona fide magic square. Switch each digit to its equivalent modulo 4. We then have another pseudomagic square, but with all digits 0, 1, 2, or 3.

$$\begin{bmatrix} 11 & 22 & 33 & 00 \\ 23 & 10 & 01 & 32 \\ 30 & 03 & 12 & 21 \\ 02 & 31 & 20 & 13 \end{bmatrix}$$

Now treat these as numbers in base 4 and find their base 10 equivalents. For example, $33_4 = 3 \cdot 4 + 3 = 15_{10}$.

$$\begin{bmatrix} 5 & 10 & 15 & 0 \\ 11 & 4 & 1 & 14 \\ 12 & 3 & 6 & 9 \\ 2 & 13 & 8 & 7 \end{bmatrix}$$

This is nearly a magic square, but the numbers run from 0 to 15 consecutively. Just add one to each entry and we have a 4×4 magic square (Fig. 8.9).

Let's go a bit further. Consider an additional 4×4 Latin square (Fig. 8.10). Notice that the three squares, A, B, and C are pairwise orthogonal Latin squares. Could there be more pairwise orthogonal Latin squares of order 4? Our next proposition says no.

$$\begin{bmatrix} 6 & 11 & 16 & 1 \\ 12 & 5 & 2 & 15 \\ 13 & 4 & 7 & 10 \\ 3 & 14 & 9 & 8 \end{bmatrix}.$$

Figure 8.9 4 × 4 magic square.

$$C = \begin{bmatrix} 1 & 2 & 3 & 4 \\ 4 & 3 & 2 & 1 \\ 2 & 1 & 4 & 3 \\ 3 & 4 & 2 & 1 \end{bmatrix}.$$

Figure 8.10 A, B, and C are pairwise orthogonal Latin squares.

Proposition 8.1: *There exist at most $n - 1$ pairwise orthogonal Latin squares of order n.*

Proof of Proposition 8.1: *For any set of pairwise orthogonal Latin squares, we may assume without loss of generality that in each square the first row is the numbers 1, 2, ..., n in consecutive order. The first rows then account for the ordered pairs (1, 1), (2, 2), ..., (n, n) in each pair of orthogonal Latin squares. Now the entry in the (2, 1) position of each Latin square must be different from one another and none can be a 1 (since we already have a 1 in the first column). There are $n - 1$ choices for the (2, 1) position, accounting for at most $n - 1$ pairwise orthogonal Latin squares.* □

By definition, $n - 1$ pairwise Latin squares of order n form a *complete set* of orthogonal Latin squares. Hence, the three squares A, B, and C comprise a complete set of order 4 orthogonal Latin squares.

Proposition 8.1 provides a limit on how many Latin squares there could be of a given order. But note that it does not guarantee that there really are any particular number of such Latin squares. A significant amount of mathematics has been developed to actually construct pairwise orthogonal Latin squares of various orders. Our next theorem is a modest step in this direction.

Proposition 8.2: *If p is prime, then there is a complete set of $p - 1$ pairwise orthogonal Latin squares of order p.*

Proof of Proposition 8.2: *We will explicitly construct a family $A_1, A_2, \ldots, A_{p-1}$ of pairwise orthogonal Latin squares of order p as follows: If $A_r(i, j)$ denotes the* *entry in the ij^{th} position of square A_r for some $r(1 \le r \le p-1)$, then let*

$$A_r(i,j) = r(i-1) + j (mod\, p). \tag{8.1}$$

Here we use the complete set of residues 1, 2, ..., p modulo p.

For any r, to see that A_r is a Latin square, suppose that two entries in the same row are identical. Then $A_r(i, j_1) = A_r(i, j_2)$ for appropriate row i and columns j_1 and j_2 where $1 \le j_1 \le j_2 \le p$. Hence by Formula 8.1,

$$r(i-1) + j_1 \equiv r(i-1) + j_2 (mod\, p)$$

from which it follows that $j_1 \equiv j_2$ (mod p). But since $1 \le j_1 \le j_2 \le p$, it must be that $j_1 = j_2$. Hence no two entries in the same row can be identical. Next suppose that two entries in the same column are identical. In particular, assume that $A_r(i_1, j) = A_r(i_2, j)$ for appropriate column j and rows i_1 and i_2 where $1 \le i_1 \le i_2 \le p$. By Formula 8.1,

$$r(i_1 - 1) + j \equiv r(i_2 - 1) + j (mod\, p).$$

This implies that $ri_1 \equiv ri_2$ (mod p). But since $1 \le r \le p-1$, the numbers r and p are relatively prime. Hence we can divide the last congruence by r without disturbing the modulus. Thus $i_1 \equiv i_2$ (mod p). But again $1 \le i_1 \le i_2 \le p$ and so $i_1 = i_2$. It follows that each of the squares are Latin squares.

Now we must show that they are pairwise orthogonal Latin squares. Suppose on the contrary that for some $r \ne s$, A_r and A_s do not form an orthogonal pair. Then there is an ordered pair of numbers (x, y) that appears twice among the pairings of entries in A_r and A_s. Let

$$x = A_r(i_1, j_1) = A_r(i_2, j_2), \text{ and}$$
$$y = A_s(i_1, j_1) = A_s(i_2, j_2)$$

where $i_1 \ne i_2$ and $j_1 \ne j_2$ (since A_r and A_s are Latin squares). The numbers x and y can be explicitly computed using Formula 8.1, so

$$r(i_1 - 1) + j_1 \equiv r(i_2 - 1) + j_2 (mod\, p), \text{ and}$$
$$s(i_1 - 1) + j_1 \equiv s(i_2 - 1) + j_2 (mod\, p).$$

Simplifying, we get

$$r(i_1 - i_2) \equiv (j_1 - j_2)(mod\, p), \text{ and}$$
$$s(i_1 - i_2) \equiv (j_1 - j_2)(mod\, p).$$

Hence

$$r(i_1 - i_2) \equiv s(i_1 - i_2)(mod\, p).$$

But since $i_1 \ne i_2$, $p \nmid (i_1 - i_2)$ and hence $r \equiv s$ (mod p). But this implies that $r = s$ since $1 \le r, s \le p-1$. This contradicts our assumption that A_r and A_s are *distinct. Therefore,* $A_1, A_2, \ldots, A_{p-1}$ *form a complete set of pairwise orthogonal Latin squares of order* p. □

Example: Mimicking the proof of Theorem 8.2, we can construct a pair of orthogonal Latin squares of order three (which is necessarily a complete set.) In this case, $A_1(i, j) = i + j - 1$ and $A_2(i, j) = 2i + j - 1$ for $1 \leq i \leq 3, 1 \leq j \leq 3$. The two Latin squares are

$$A_1 = \begin{bmatrix} 1 & 2 & 3 \\ 2 & 3 & 1 \\ 3 & 1 & 2 \end{bmatrix}, A_2 = \begin{bmatrix} 1 & 2 & 3 \\ 3 & 1 & 2 \\ 2 & 3 & 1 \end{bmatrix}.$$

Using a little finite field theory, Theorem 8.2 can be extended to show that for $n = p^a$ where p is prime and $a \geq 1$ that there is a complete set of pairwise orthogonal Latin squares of order n. But for several other values of n, large sets of pairwise orthogonal Latin squares are harder to come by.

Euler discovered algorithms to generate pairs of orthogonal Latin squares of order n for many values of n except for those $n \equiv 2 \pmod 4$. He was convinced that there were no pairs of orthogonal Latin squares of order 6 (confirmed by Gaston Tarry in 1901 through an exhausting and exhaustive search) and used this result as a basis for his Thirty-six Officers Problem. The problem states that there are 36 military officers of six different ranks and from six different regiments. Each combination of rank and regiment is represented among the 36 officers. Line up the officers in six rows of six so that each rank and file has an officer of each rank and every regiment. Of course, by Euler's previous result such an arrangement is impossible.

In 1782, Euler conjectured that there are no orthogonal Latin squares of order $4k+2$ for any $k \geq 1$. Unfortunately, Euler was as wrong about this as Fermat had been about Fermat primes. In 1959, E.T. Parker, R.C. Bose, and S.S. Shrikhande constructed an order 10 pair of orthogonal Latin squares. They then showed how to make similar constructions for all $4k+2$ for any $k \geq 2$. Since then, many order 10 orthogonal pairs have been constructed. However, no one has been able to construct a complete set of order 10 orthogonal Latin squares. In fact, no one has even constructed a set of three mutually orthogonal Latin squares of order 10.

We complete this chapter with an example of a pair of orthogonal Latin squares of order 10.

$$\begin{bmatrix} 1 & 8 & 9 & 10 & 2 & 4 & 6 & 3 & 5 & 7 \\ 7 & 2 & 8 & 9 & 10 & 3 & 5 & 4 & 6 & 1 \\ 6 & 1 & 3 & 8 & 9 & 10 & 4 & 5 & 7 & 2 \\ 5 & 7 & 2 & 4 & 8 & 9 & 10 & 6 & 1 & 3 \\ 10 & 6 & 1 & 3 & 5 & 8 & 9 & 7 & 2 & 4 \\ 9 & 10 & 7 & 2 & 4 & 6 & 8 & 1 & 3 & 5 \\ 8 & 9 & 10 & 1 & 3 & 5 & 7 & 2 & 4 & 6 \\ 2 & 3 & 4 & 5 & 6 & 7 & 1 & 8 & 9 & 10 \\ 3 & 4 & 5 & 6 & 7 & 1 & 2 & 10 & 8 & 9 \\ 4 & 5 & 6 & 7 & 1 & 2 & 3 & 9 & 10 & 8 \end{bmatrix}$$

$$\begin{bmatrix} 1 & 7 & 6 & 5 & 10 & 9 & 8 & 2 & 3 & 4 \\ 8 & 2 & 1 & 7 & 6 & 10 & 9 & 3 & 4 & 5 \\ 9 & 8 & 3 & 2 & 1 & 7 & 10 & 4 & 5 & 6 \\ 10 & 9 & 8 & 4 & 3 & 2 & 1 & 5 & 6 & 7 \\ 2 & 10 & 9 & 8 & 5 & 4 & 3 & 6 & 7 & 1 \\ 4 & 3 & 10 & 9 & 8 & 6 & 5 & 7 & 1 & 2 \\ 6 & 5 & 4 & 10 & 9 & 8 & 7 & 1 & 2 & 3 \\ 3 & 4 & 5 & 6 & 7 & 1 & 2 & 8 & 10 & 9 \\ 5 & 6 & 7 & 1 & 2 & 3 & 4 & 10 & 9 & 8 \\ 7 & 1 & 2 & 3 & 4 & 5 & 6 & 9 & 8 & 10 \end{bmatrix}$$

WORTH CONSIDERING

1. Use De la Loubère's algorithm to generate a 7×7 magic square.

2. Investigate the existence of a knight's tour on $n \times n$ chessboards for $n = 3, 4,$ and 5.

3. Show that there are no orthogonal Latin squares of order 2.

4. Find all pairs of orthogonal Latin squares of order 3.

5. **(a)** Find a pair of orthogonal Latin squares of order 5.

(b) Can you find a complete set of orthogonal Latin squares of order 5?

6. Explain how a 4×4 Latin square might be useful in planning the tire rotation for your car.

7. Arrange the numbers 1 to 27 in a $3 \times 3 \times 3$ magic cube so that each row, column, and "pillar" adds up to 42.

8. Golf Problem (Steve Abbott): Arrange for a group of 16 golfers to golf for five days, each day in four groups of four, so that each golfer plays exactly once with every other golfer.

9. An $r \times n$ Latin rectangle has r rows and n columns with the numbers $1, 2, \ldots, n$ in each row with no number repeated in any column. The number of $2 \times n$ Latin rectangles is known as the *derangement number* of order n, denoted by D_n. Calculate D_n for $n = 2$, 3, 4, and 5.

10. **(a)** Let $L_n =$ the total number of Latin squares of order n. Show that $n!(n-1)!|L_n$ by considering those Latin squares having first row and first column in consecutive order $1, 2, \ldots, n$.

(b) Let $l_n = L_n/n!(n-1)!$. Find l_n for $1 \leq n \leq 4$. (For the record, $l_5 = 56, l_6 = 9,408$, and $l_7 = 16,942,080$.)

9 Mathematical Variations on Rolling Dice

In addition to their recreational value in games of chance, dice play a role (or is it roll?) in several interesting mathematical problems. In this chapter, we present a few nice problems involving dice. As usual, we start with easier problems and develop the requisite mathematics as we go. Please do make the effort. The results are quite intriguing!

An ordinary die consists of a cube (regular hexahedron) with the numbers one through six inclusive inscribed on the sides. The ordering of the numbers is fixed so that the sum of opposite sides is always seven. To view a die in two dimensions, we can flatten out the surface of the die and represent the sides as shown in Figure 9.1.

If we roll two dice and note their sum, there are 11 different outcomes. However, the probability of the outcomes vary depending on the number of ways that a given sum can be created. For example, the most likely single outcome is a sum of seven which can occur as $1 + 6, 2 + 5, 3 + 4, 4 + 3, 5 + 2$, or $6 + 1$. It's this variation in the likelihood of different outcomes which adds interest and complexity to games of chance involving dice. Below we tabulate the probability of obtaining each of the possible sums for a pair of fair dice (Table 9.1).

Let $P(r)$ denote the probability that the sum of the two dice is r. Two basic facts about probability distributions are the following:

1. $P(r) \geq 0$ for all $r = 2, 3, \ldots, 12$.

2. $\sum_{r=2}^{12} P(r) = 1$.

Example #1: *Can we weight two dice so that all possible outcomes (of sums) are equally likely?*

We have already handled the analogous problem with two coins (Problem #4, Chapter 2), but this is a good warm-up and we will present a slightly different

Mathematical Journeys, by Peter D. Schumer
ISBN 0-471-22066-3

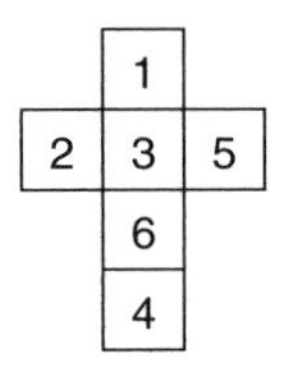

Figure 9.1 Ordinary die flattened.

TABLE 9.1 Likelihood of sums for a pair of dice

Outcome	2	3	4	5	6	7	8	9	10	11	12
Probability	$\frac{1}{36}$	$\frac{2}{36}$	$\frac{3}{36}$	$\frac{4}{36}$	$\frac{5}{36}$	$\frac{6}{36}$	$\frac{5}{36}$	$\frac{4}{36}$	$\frac{3}{36}$	$\frac{2}{36}$	$\frac{1}{36}$

solution than the one given previously. We first make a small algebraic observation:

For any real numbers x and y with $xy \neq 0, x^2 + y^2 > xy$.

To see this, note that $(x - y)^2 = x^2 - 2xy + y^2 \geq 0$, and hence $x^2 + y^2 \geq 2xy$. But if $xy > 0$, then $2xy > xy$ and $x^2 + y^2 > xy$. If $xy < 0$, then again $x^2 + y^2 > xy$ since the left-hand side is positive.

Solution to Problem #1: *Let $p_1, p_2, \ldots, p_6$ denote the probabilities of rolling a $1, 2, \ldots,$ or 6 on the first die, respectively. Similarly, let $P_1, P_2, \ldots, P_6$ denote the probabilities of rolling a $1, 2, \ldots,$ or 6 on the second die. Suppose that all sums are equally likely. Then the probability of any particular sum is $\frac{1}{11}$. In particular, $P(2) = p_1 \cdot P_1 = \frac{1}{11}$ and thus $p_1 = \frac{1}{11P_1}$. Analogously, $P(12) = p_6 \cdot P_6 = \frac{1}{11}$ and thus $p_6 = \frac{1}{11P_6}$. Now*

$$\begin{aligned} P(7) &= p_1P_6 + p_2P_5 + p_3P_4 + p_4P_3 + p_5P_2 + p_6P_1 \\ &\geq p_1P_6 + p_6P_1 \\ &= \frac{P_6}{11P_1} + \frac{P_1}{11P_6} \\ &= \frac{1}{11}\left(\frac{P_6}{P_1} + \frac{P_1}{P_6}\right) \\ &= \frac{1}{11}\left(\frac{P_1^2+P_6^2}{P_1P_6}\right). \end{aligned}$$

But by letting $x = P_1$ and $y = P_6$ in our algebraic observation above, we deduce that $P(7) > \frac{1}{11}$ since the fraction in the parentheses above is greater than one.

Hence no equiprobable-sum weighting of the two dice is possible. □

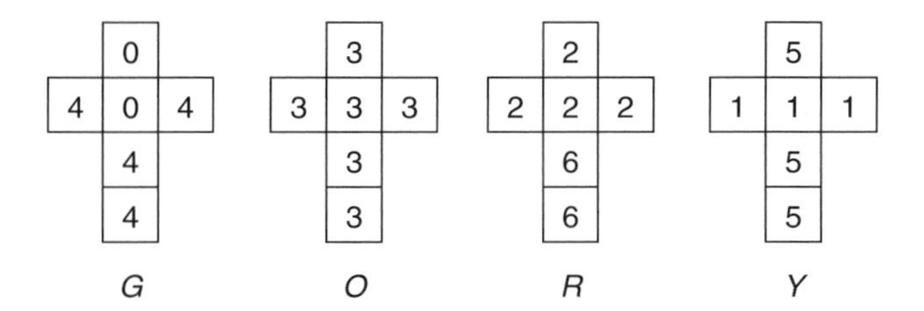

Figure 9.2 Efron's dice.

Example #2: *(Efron's Dice) The next example is a fascinating set of four non-standard dice created by the Stanford statistician Bradley Efron in 1970. These four funny dice have a surprising nontransitive property that we will describe fully. The four dice are colored green (G), orange (O), red (R), and yellow (Y) (Fig. 9.2).*

Many mathematical systems satisfy the transitive property. For example, if x, y, and z are three real numbers and $x > y$ and $y > z$, then it must follow that $x > z$. The transitive property is so nearly ubiquitous that we tend to take for granted that it always applies. However, here is one setting where we would be mistaken to assume transitivity.

With the four dice just described, two players play the following game: Each chooses a different one of the four dice, rolls it, and the winner is the one with the larger number. Ties cannot occur since no two dice share a common number. Let's see what happens with certain choices of dice. (We refer to the players by the color of the die they choose.)

If green (G) plays against orange (O), then G wins as long as a 4 is rolled and loses if a 0 is rolled. Hence the probability that G beats O is $\frac{4}{6}$ or $\frac{2}{3}$. We denote this by $P(G > O) = \frac{2}{3}$. Similarly, $P(O > R) = \frac{2}{3}$ since a 3 will beat any of the four 2's, but loses to either of the 6's. What happens when red plays yellow? Red has $\frac{1}{3}$ chance of rolling a 6, in which case red wins. If red rolls a 2 (which happens with probability $\frac{2}{3}$), then red has a $\frac{3}{6} = \frac{1}{2}$ chance of beating yellow (depending on whether Y rolls a 1 or a 5). So $P(R > Y) = \frac{1}{3} + \frac{1}{2} \cdot \frac{2}{3} = \frac{2}{3}$. And now, how about yellow versus green? Yellow has a $\frac{1}{2}$ chance of rolling a 5, in which case yellow has a guaranteed win. If yellow rolls a 1 (which occurs with probability $\frac{1}{2}$), then yellow has two chances in six or probability $\frac{1}{3}$ of winning. Hence $P(Y > G) = \frac{1}{2} + \frac{1}{2} \cdot \frac{1}{3} = \frac{2}{3}$. Summarizing our results,

$$P(G > O) = \tfrac{2}{3}, P(O > R) = \tfrac{2}{3}, P(R > Y) = \tfrac{2}{3}, \text{ and } P(Y > G) = \tfrac{2}{3}.$$

The outcomes of these four dice is highly nontransitive. In practice, you could politely let your opponent choose first. Once your opponent has chosen a die, you always have the opportunity of choosing a remaining one which gives you a significant chance of winning. With a frustrated opponent the situation could get a bit gory, which is why I chose the colors GORY to help you remember which die to choose.

Example #3: *Is there a way to renumber a pair of ordinary dice so that the probability distribution for the sum is identical to that for a normal pair of dice?*

By way of background we need to introduce a very important topic, namely that of *generating functions*. Rather than thinking of the outcomes of a die's roll as being a 1, 2, 3, 4, 5, or 6 we now think of rolling an x^1, x^2, x^3, x^4, x^5, or x^6 respectively. So the generating function for a single normal die is the function $f(x) = 1x^1+1x^2+1x^3+1x^4+1x^5+1x^6$. In each term, the exponent represents a possible outcome and the coefficient indicates the number of sides with that value.

If we roll two dice, then each die has its own generating function, $f_1(x) = 1x^1+1x^2+1x^3+1x^4+1x^5+1x^6$ and $f_2(x) = 1x^1+1x^2+1x^3+1x^4+1x^5+1x^6$. The sum of two dice is represented by the product $f_1(x) \cdot f_2(x)$. By the law of exponents, add exponents when multiplying monomials with the same base, the coefficient of each term gives the number of ways that sum can occur. In particular,

$$f_1(x) \cdot f_2(x) = 1x^2 + 2x^3 + 3x^4 + 4x^5 + 5x^6 + 6x^7 + 5x^8 + 4x^9 + 3x^{10} + 2x^{11} + 1x^{12}.$$

To answer the question of whether a standard pair of dice can be renumbered and still have the same probability distribution, it is necessary and sufficient to find two other generating functions $g_1(x)$ and $g_2(x)$ such that:

1. $g_1(x) \cdot g_2(x) = f_1(x) \cdot f_2(x)$,
2. $g_1(1) = 6 = g_2(1)$, and
3. g_1 and g_2 have no constant terms.

Condition (1) ensures that the sum of a roll of the renumbered dice behave the same way as do our standard dice. Condition (2) checks that the new dice are still six-sided. The last condition says that on any given roll, some side must come up. In other words, there is no chance that either die ends up balanced on an edge or a corner. In addition, we don't have any blank side, that is, a side with value zero.

To proceed, factor $f_1(x) \cdot f_2(x)$ as far as possible into a product of irreducible polynomials. Each die factors identically as

$$\begin{aligned} x + x^2 + x^3 + x^4 + x^5 + x^6 &= x(1 + x + x^2 + x^3 + x^4 + x^5) \\ &= x\left(\frac{x^6 - 1}{x - 1}\right) \\ &= \frac{x(x^3 + 1)(x^3 - 1)}{x - 1} \\ &= x(x^3 + 1)(x^2 + x + 1) \\ &= x(x + 1)(x^2 - x + 1)(x^2 + x + 1). \end{aligned}$$

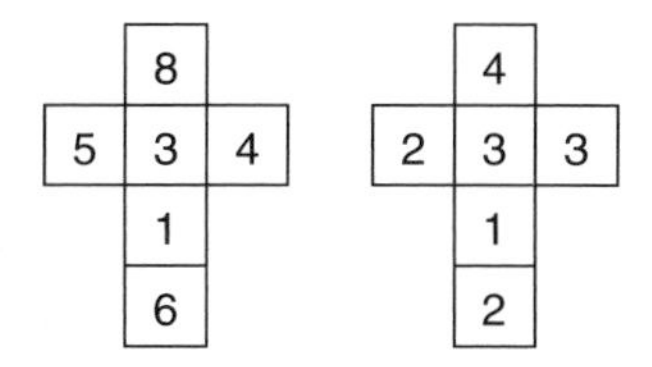

Figure 9.3 Odd dice that behave like normal dice.

Thus

$$f_1(x) \cdot f_2(x) = x^2(x+1)^2(x^2-x+1)^2(x^2+x+1)^2 \tag{9.1}$$

By condition (1), together $g_1(x)$ and $g_2(x)$ must comprise all the factors in the right-hand side of Equation 9.1. By condition (2) both $g_1(x)$ and $g_2(x)$ must contain the factors $x+1$ and x^2+x+1 since there is no other way to make $g_1(1) = 6 = g_2(1)$. By condition (3) each generating function must contain a factor of x. But now all that remains is to assign the remaining $(x^2-x+1)^2$. However, if we assign one copy of this polynomial to each of g_1 and g_2, then $g_1 = f_1$ and $g_2 = f_2$. Hence there is only one other possibility, namely assign $(x^2-x+1)^2$ to one of the g's , say g_1. Therefore, $g_1(x) = x(x^2+x+1)(x^2-x+1)^2(x+1)$ and $g_2(x) = x(x^2+x+1)(x+1)$. Expanding the products we obtain

$$g_1(x) = x^8 + x^6 + x^5 + x^4 + x^3 + x \text{ and } g_2(x) = x^4 + 2x^3 + 2x^2 + x.$$

The resulting dice are shown in Figure 9.3.

Furthermore, if we are willing to have dice which are not cube-shaped, then there are other possibilities created by relaxing condition (2). For example, we can define $h_1(x) = x(x+1)^2$ and $h_2(x) = x(x^2-x+1)^2(x^2+x+1)^2$. In this case, expanding produces

$$h_1(x) = x^3 + 2x^2 + x \text{ and } h_2(x) = x^9 + 2x^7 + 3x^5 + 2x^3 + x.$$

This produces a four-sided die with values (1, 2, 2, 3) and a nine-sided die with values (1, 3, 3, 5, 5, 5, 7, 7, 9). A tetrahedron can be used to provide a regular four-sided polyhedron, but there is no regular polyhedron with nine sides. What can be done? Make a cylinder with each face a regular nine-sided polygon (nonagon). More generally, if an n-sided die is needed, a cylinder with each face a regular n-sided polygon works. Each roll, however, is the number which faces *down* once the roll is complete. Which side is facing up may be ambiguous. There is another way to assign two polynomials that will result in a four-sided and a nine-sided die with the same product as $h_1(x) \cdot h_2(x)$, but we will leave it for the exercises.

We complete this chapter with a cute problem involving a five-sided and a six-sided die.

Example #4: *Roll two dice, a standard six-sided die numbered 1 through 6 and a "regular" five-sided die numbered 1 through 5. What's the probability that the number on the six-sided die is larger than that on the five-sided die?*

Solution: *Roll the dice and let R be the number on the six-sided die and r be the number on the five-sided die. The complementary number to R (on the flip side of the six-sided die) is the number $7 - R$. Note that the complementary numbers are just 1 through 6 again, this time written in reverse order. Analogously, define the complementary number to r to be $6 - r$. Here again the complementary numbers comprise 1 through 5 appearing in reverse order.*

We must find the probability that $R > r$. It must be the case that either $R - r > 0$ or that $R - r \leq 0$. These cases are mutually exclusive (they can't both happen) and exhaustive (nothing else can happen). Since the numbered faces are all whole numbers we can rewrite the two possibilities as either $R - r > 0$ or $R - r < 1$. The first possibility can be rewritten as $R > r$ and the second (after adding 6 to both sides and rearranging) as $7 - R > 6 - r$. Hence either the number on the six-sided die is larger than that on the five-sided die or the complementary number on the six-sided die is larger than the complementary number on the five-sided die. These two possibilities are exhaustive and, by symmetry, they are equally likely as well. (We could have defined each roll by its complementary number in the first place.) Hence each has probability one-half and so $P(R > r) = \frac{1}{2}$.

WORTH CONSIDERING

1. Determine how to weight two dice so that the probability of rolling a sum of 2, 3, 4, 5, or 6 is $\frac{1}{16}$ in all cases.
2. With Efron's dice in Example #2, which one would you choose if the goal is to have the highest total after five rolls of a single die?
3. In Example #2, what is the chance of winning if all competitions are best-of-three?
4. Use generating functions to determine the number of ways three dice can sum to ten.
5. Find another way besides (1, 2, 2, 3) and (1, 3, 3, 5, 5, 5, 7, 7, 9) to assign positive integers to a four-sided and a nine-sided die that has the same probability distribution as a pair of standard dice.
6. How would our analysis of Example #3 be modified if we did allow for sides with value zero?
7. (R.M. Shortt, S.G. Landry, L.C. Robertson, 1988) Consider the two ten-sided *weighted* dice given by the generating functions

   $f_1(x) = x + \frac{1+\sqrt{5}}{2}x^3 + x^5 + x^6 + \frac{1+\sqrt{5}}{2}x^8 + x^{10}$ and $f_2(x) = x + 2x^2 + \frac{5-\sqrt{5}}{2}x^3 + (3-\sqrt{5})x^4 + (4-\sqrt{5})x^5 + (4-\sqrt{5})x^6 + (3-\sqrt{5})x^7 + \frac{5-\sqrt{5}}{2}x^8 + 2x^9 + x^{10}$. Compute $f_1(x) \cdot f_2(x)$ and interpret the result.

8. Roll a five-sided die and a six-sided die. What's the probability that the six-sided die has a smaller roll? What about the two dice having the same roll?

9. **(a)** Roll a six-sided die and a seven-sided die. What's the probability that the seven-sided die has the larger roll?

(b) Roll a four-sided die and a six-sided die. What's the probability that the six-sided die has the larger roll?

10. Design three six-sided dice having nontransitive pairwise outcomes.

10 Pizza Slicing, Map Coloring, Pointillism, and Jack-in-the-Box

In his book, *The Teaching of Geometry* (1911), D.E. Smith said, "Geometry is a mountain. Vigor is needed for its ascent. The views all along the paths are magnificent. The effort of climbing is stimulating. A guide who points out the beauties, the grandeur, and the special places of interest commands the admiration of his group of pilgrims." Even though this quote might be a bit over the top, I invite you to be your own guide as we discuss four different problems, each with a different geometric view.

Problem #1 (Pizza Slicing Problem): *How many regions can the plane be separated into with n straight lines?*

We can think of the plane as a gigantic pizza and the straight lines as the cuts made across the pie. To get a feel for this problem, make some sketches with small values of n and see if a pattern emerges. This problem appears so naturally; if you're like me you've probably doodled with this problem on place mats and scraps of paper since childhood. Its mathematical formulation was stated and solved by the great synthetic geometer, Jakob Steiner (1796–1863), in an article in Crelle's Journal (1826). Steiner was Swiss, but was educated in Heidelberg and spent most of his career as a professor at the University of Berlin. It may be of interest to note that the man who created so many beautiful geometric theorems didn't learn to read or write until he was 14 years old. To mention just one of his accomplishments: The Danish mathematician Georg Mohr (1640–1697) had shown that all Euclidean constructions (those involving straightedge and compass) could actually be accomplished with compass alone, as long as we consider a line to exist once two of its points are so constructed. In the other direction, Steiner showed that all Euclidean constructions can be accomplished with one fixed circle plus a straightedge, that is, with a straightedge and a compass that's stuck!

Problem #2 (Map Coloring Problem): *Slice up the plane with any number of straight lines. How many colors are required so that adjoining regions have different colors?*

Mathematical Journeys, by Peter D. Schumer
ISBN 0-471-22066-3

Problem #2 begins where Problem #1 left off. Once we have sliced up the plane, we now want to color the "map" that we've created.

For comparison's sake, a much more difficult and appropriately celebrated problem is the question of how many colors are needed to color any map in the plane so that adjoining regions have different colors. Boundaries need no longer be just straight lines. This significant problem confounded amateur and professional mathematicians alike from the mid-1800's until 1976, when Kenneth Appel and Wolfgang Haken of the University of Illinois, assisted by an enormous amount of computer time (about 1,200 hours), established that four colors suffice. The fact that five colors suffice was established by P.J. Heawood in 1890 and he didn't need any computational assistance, but the step from five colors to four colors is a large one indeed.

Problem #3 (Pointillism Problem): *If every point of the plane is colored either red, green, or blue, are there necessarily at least two points of the same color one unit apart?*

This problem reminds me of the paintings of Georges Seurat and hence the allusion to pointillism. Notice that we have not defined what our units are. Interestingly, it makes no difference. To get a feel for this type of problem, consider the situation where every point of the plane is colored either red or green. Would there be two points of the same color one unit apart? The answer is "yes" as can be easily seen by considering any equilateral triangle in the plane with sides of unit length. If one vertex is red and another green, what color could the third vertex be? If it is red, then there are two points one unit apart both of which are red. If it is green, then there are two green points one unit apart. Either way, there are two points of the same color a unit apart.

Problem #4 (Jack-in-the-Box Problem): *Consider Figure 10.1, where four unit circles in the plane bound a smaller circle placed within them so that the inner circle is tangent to the four unit circles. The inner circle lies well within the black square containing the four unit circles. In three-dimensions, eight unit balls surround a ball that remains within the cube containing the eight unit balls. What happens in higher dimensions? Does the "inner ball" remain within the outer hypercube or does it somehow pop out like a wind-up jack-in-the-box?*

Solution to Problem #1: *Instead of working in the entire plane, we limit our view to a large circle. Let L_n = the maximum number of regions created by drawing n straight lines. Certainly the lines must be in general position, that is, no three lines intersect at a point and no two lines are parallel to each other. Figure 10.2 shows the situation with five straight lines. In this case, the circle is separated into 16 distinct regions. In Table 10.1 we chart the growth of L_n for n from 0 to 5. Analyzing Table 10.1, we see for $1 \leq n \leq 5$ that $L_n = L_{n-1} + n$. If we can show that this pattern continues indefinitely, then it would follow that for all $n \geq 1$:*

$$L_n = 1 + 1 + 2 + \ldots + n = 1 + \frac{n(n+1)}{2} = \frac{n^2+n+2}{2}.$$

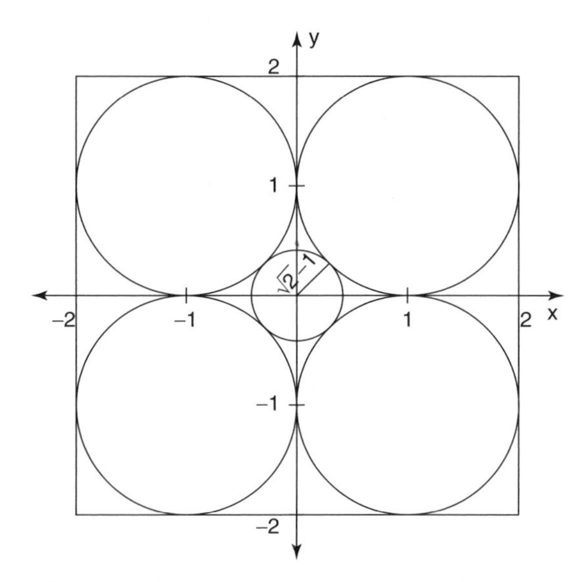

Figure 10.1 Circle bounded by four unit circles.

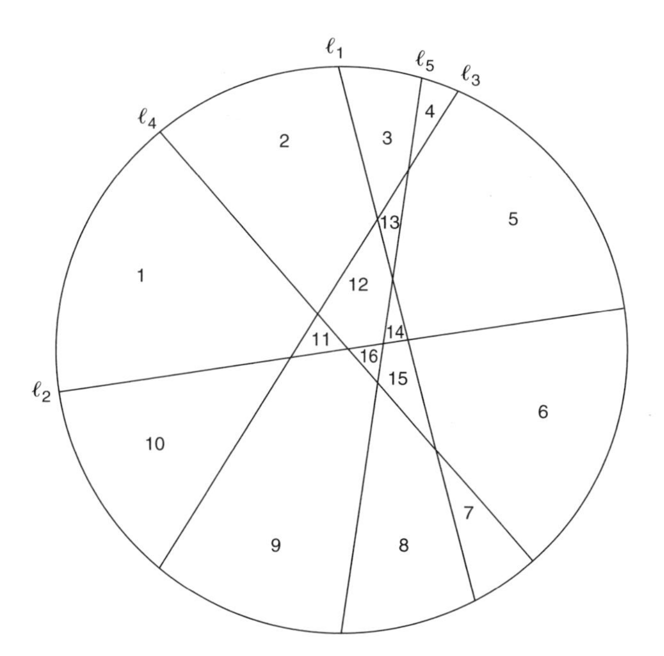

Figure 10.2 Five lines separate the plane into sixteen regions.

We now show that $L_n = L_{n-1} + n$ *for all* $n \geq 1$ *by induction on* n.

For $n = 1, L_1 = 2 = L_0 + 1$. *Now assume the result holds up to* $n - 1$ *lines. Since a line is determined by two points, in order to place the* n^{th} *line so as to maximize the number of new regions, it suffices to choose two points on*

TABLE 10.1 Number of regions created by n straight lines

n	0	1	2	3	4	5
L_n	1	2	4	7	11	16

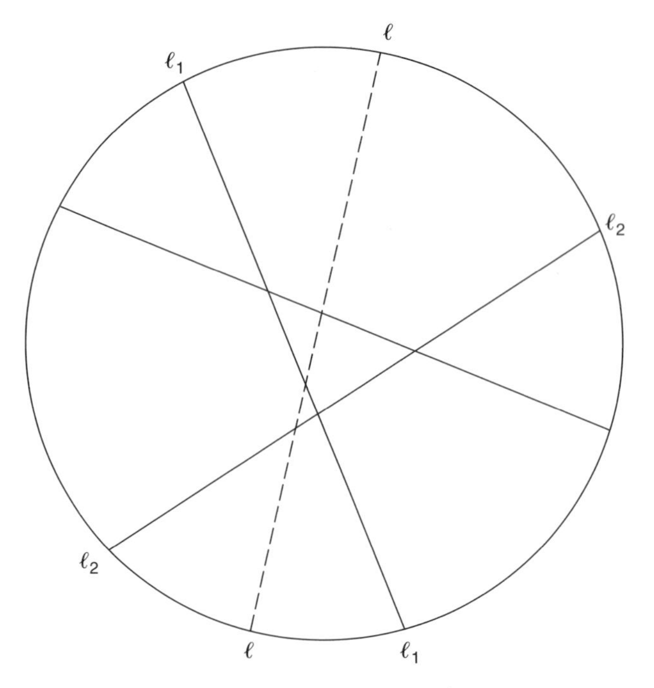

Figure 10.3 Placement of nth line.

the circumference of the circle that lie directly between the same two lines (as in Fig. 10.3), thus avoiding parallelism. Furthermore, we must make sure that the new line does not go through any previous points of intersection. This can be easily done since avoiding such points only reduces our infinitely many choices on the circumference by a finite number.

By construction, the new line must intersect each and every one of the previous lines exactly once (since they are straight lines). But the new line divides previous regions into two parts for each consecutive pair of lines that it meets. Counting the circumference of the circle among these lines, the new line creates n new regions. Hence $L_n = L_{n-1} + n$ *as desired.* □

Solution to Problem #2: *The answer is two colors. To get a feel for this problem, look at Figure 10.4, where we have sliced up a large circle with four slices (line segments) and then colored the resulting regions with just two shades (dark and*

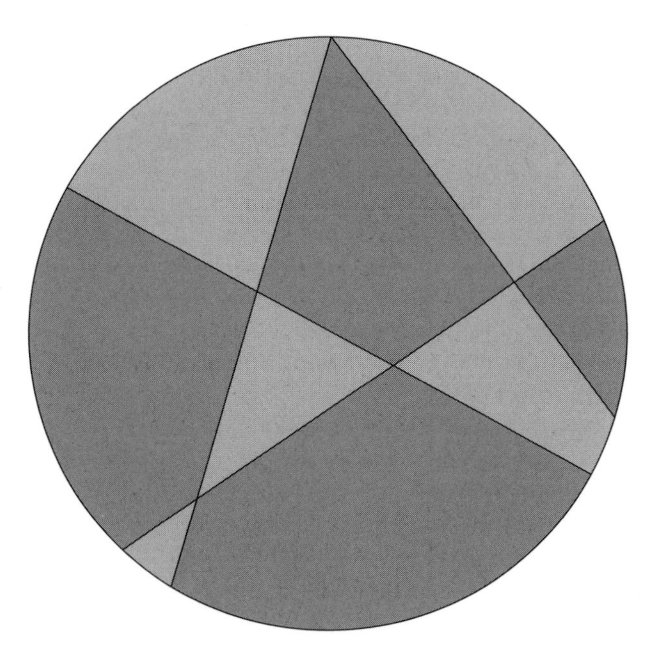

Figure 10.4 Four slices can be two-colored.

light). I invite you to check the cases with one, two, or three lines to check that two colors suffice in those cases.

Assume that for any map with n lines that two colors are all that is needed. Now consider Figure 10.5 with n + 1 lines. If the dotted line l is removed, there remains n lines. By our inductive assumption, the remaining figure can be two-colored as shown. If we now add in the dotted line, it separates the diagram into two parts—call them top and bottom. Keep the top colored as before, but switch all the colors in the bottom region (Fig. 10.6). We have now two-colored the new map. To see why this must work, we consider three possible cases in turn:

1. *Regions that border line l used to be one color on each side, but now are separated into two colors at the boundary of l.*
2. *Top regions not bordering line l have the same color as before, but these regions only border other top regions and our previous map was properly two-colored.*
3. *Similarly, bottom regions not bordering line l have all colors flipped. But these regions only border other bottom regions and the previous two-coloring is preserved by switching all such regions.* □

Solution to Problem #3: *Assume that the answer is "no," namely that there is some way to color the plane with no two points of the same color exactly one unit apart. Hence, any unit equilateral triangle will have vertices of all three colors. Consider such a triangle, △RGB as in Figure 10.7. If we use line segment GB*

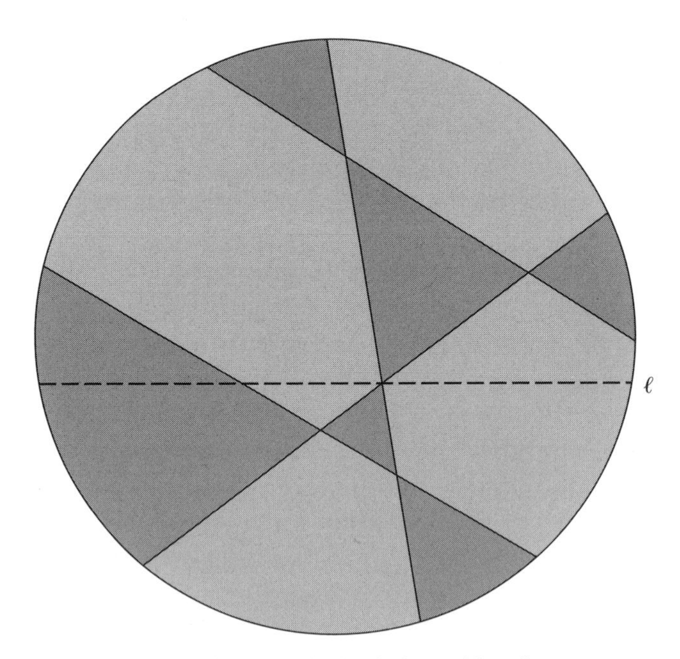

Figure 10.5 Original coloring with n lines.

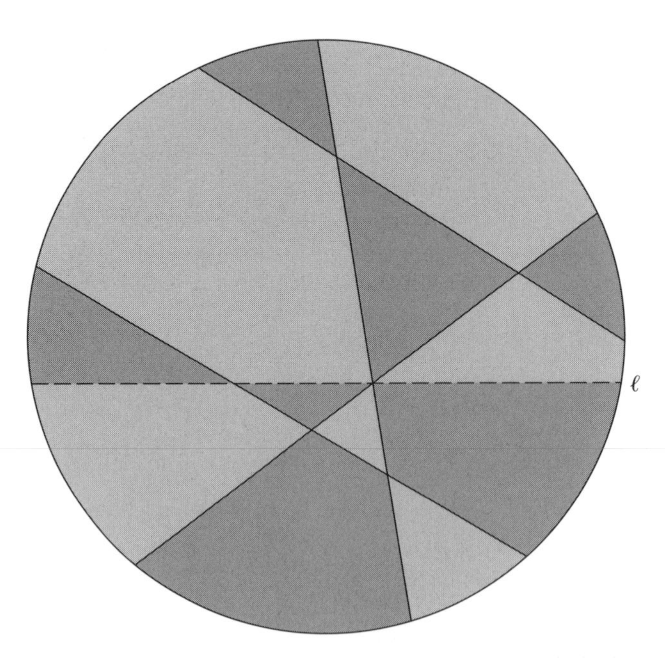

Figure 10.6 New coloring with bottom color switched.

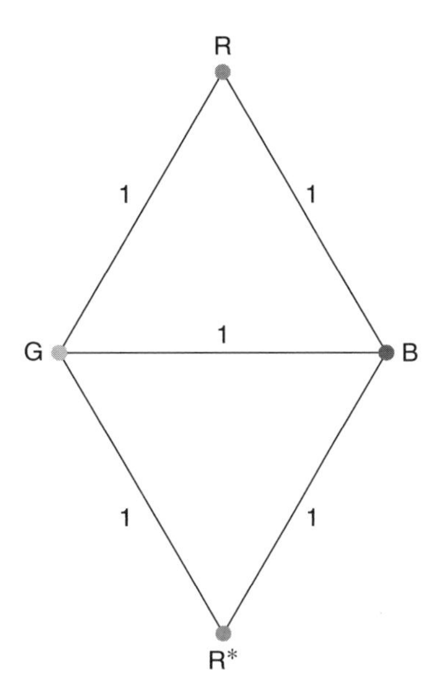

Figure 10.7 Two triangles with vertices red, green, and blue.

as one side of a new equilateral triangle, we can form another unit equilateral triangle, $\triangle R^*GB$ with vertices colored red, green, and blue.

Now "bolt down" vertex R and spin the diamond-shaped region $RGBR^*$ counterclockwise as a rigid motion with R as the pivot point. Continue until R^* moves to a point R_1 that is one unit away. Call the points where vertices G and B end up G_1 and B_1, respectively. The situation is diagrammed in Figure 10.8. The point B_1 must be either blue or green since B_1 is one unit away from the red point R. Analogously, the point G_1 must be either green or blue (whichever color B_1 is not). In either event, R_1 must be red since it is one unit from both B_1 and G_1. But then both R^* and R_1 are red and one unit apart. □

At this point, the following question is a natural one. Is there a coloring of the plane with any finite number of colors such that no two points a unit apart have the same color? If so, by Problem #3, the answer must be greater than three. The answer is "yes" and it can be shown that seven colors suffice, as in Figure 10.9. The key idea is to tile the plane with regular hexagons of side length $\frac{2}{5}$ and then color them with seven colors as shown. It can then be shown that no two points of distance d with $\frac{4}{5} < d < \frac{\sqrt{28}}{5}$ have the same color. Since $\frac{4}{5} < 1 < \frac{\sqrt{28}}{5}$, this seven-coloring of the plane does the trick.

The next question, known as the chromatic number problem for the plane, is still unresolved. What is the minimum number of colors needed to paint the plane so that no two points at unit distance have the same color? The answer conceivably could be four, five, six, or seven. It's time to get out your palette and start painting!

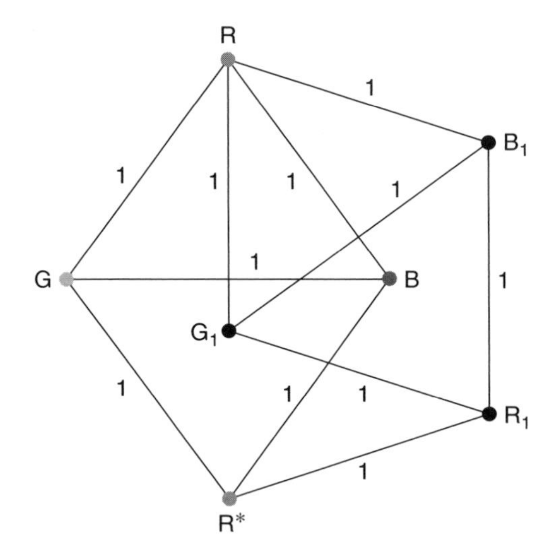

Figure 10.8 Rhombus RGR*B pivoted to RG, R, B.

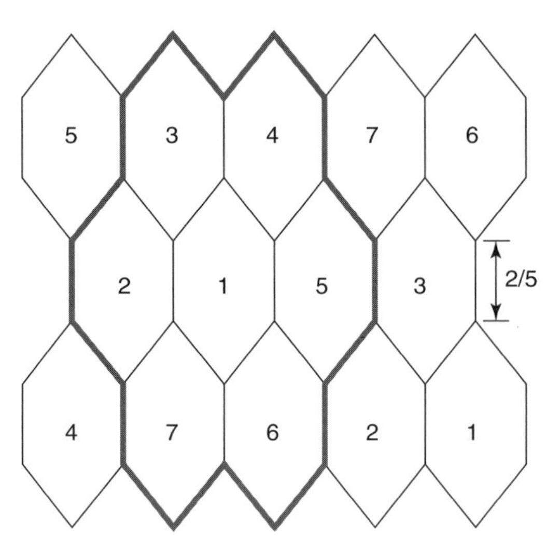

Figure 10.9 Tessellation of the plane with black-bordered 18-sided polygon of side lengths $\frac{2}{5}$.

Solution to Problem #4: *Let $\mathbb{R}^n$ represent n-dimensional Euclidean space and let r_n denote the radius of the inner n-dimensional ball contained within the outer unit balls. Figure 10.1 shows the situation in the plane when $n = 2$. Here the four unit outer circles have centers at the points (1, 1), (1, −1), (−1 , 1). and (−1 , −1). They are contained within the square bounded by the lines $x = 2$, $x = -2$, $y = 2$, and $y = -2$ of side length four. Since the center of the outer balls lie at a distance of $\sqrt{2}$ from the origin and each has radius one, being tangent to the four outer*

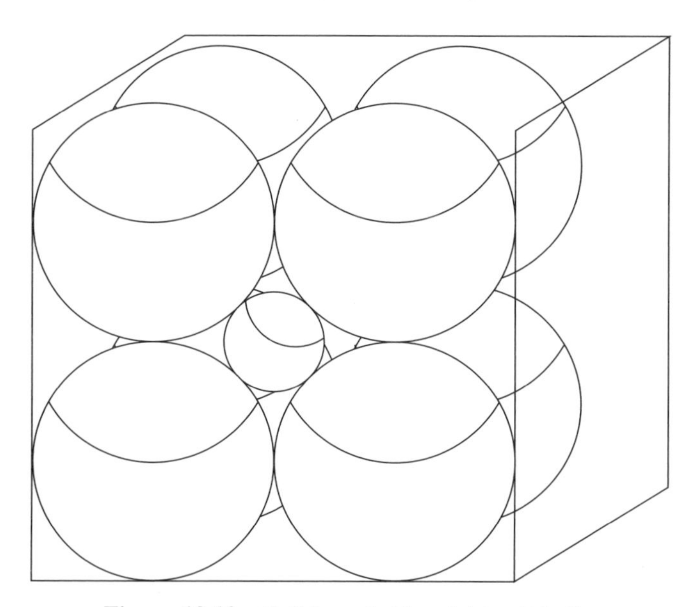

Figure 10.10 Ball bounded by eight unit balls.

circles, the inner circle has radius $r_2 = \sqrt{2} - 1$ and is certainly bounded within the outer square.

In $\mathbb{R}^3$ the eight outer balls are centered at $(\pm 1, \pm 1, \pm 1)$. You can make a model of the situation with eight tennis or lacrosse balls with a golf ball squeezed in the middle (Fig. 10.10). In this case, the outer cube is bounded by the planes $x = \pm 2$, $y = \pm 2$, and $z = \pm 2$. The outer balls have radius one, and hence the inner ball has radius $r_3 = \sqrt{3} - 1$, keeping it well within the containing cube.

By analogy, the general situation is easily described. In $\mathbb{R}^n$ there are 2^n unit outer balls centered at $(\pm 1, \pm 1, \ldots, \pm 1)$. They are contained within a hypercube bounded by the 2^n hyperplanes $x_1 = \pm 2, x_2 = \pm 2, \ldots,$ and $x_n = \pm 2$. The inner sphere has radius $r_n = \sqrt{n} - 1$. Notice for $n = 9$ that $r_9 = \sqrt{9} - 1 = 2$, and so the "inner" ball just touches the hypercube on all its faces. When $n \geq 10$, the inner ball actually pokes out of the hypercube. Pop goes the weasel! So in dimension ten, although the round inner ball is surrounded by over a thousand unit balls, it somehow manages to squeeze out beyond them. If this seems pretty weird, it may be because all the experiences in our local world involve substantially fewer dimensions. Even so, the result is both perplexing and wonderful. Don't you agree? □

WORTH CONSIDERING

1. Consider n pizza slices in general position as in Problem #1. How many of the regions border the circumference of the circle? In other words, how many such pizza slices have a crust?

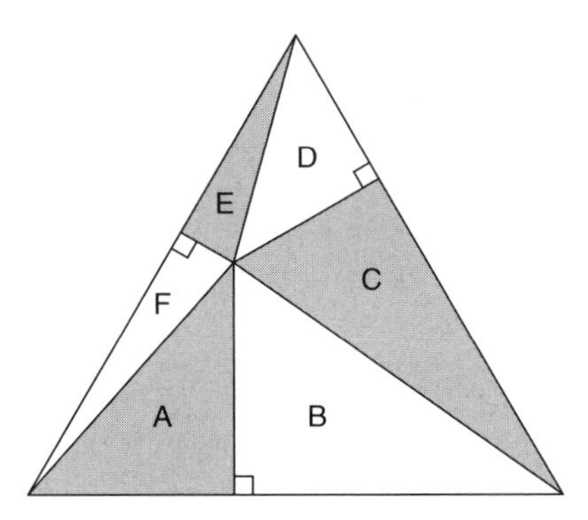

Figure 10.11 Equilateral triangle with six internal slices.

2. If every point on a straight line is colored either red or blue, must there be two points one unit apart having the same color?
3. **(a)** If all points in three-space are colored either blue, green, or red, must there be two points one unit apart that are the same color?

   (b) What about four colors in three-space?
4. (Problem from *Quantum*, January, 1990) Consider a pizza in the shape of an equilateral triangle. Pick any point inside it and make six slices by cutting from the chosen point to each of the three vertices and from the chosen point perpendicularly to each side (Fig. 10.11). Show that if two people consume the pizza by eating alternating slices, each gets exactly half the pie.
5. (Green Chicken Contest, 1986) Suppose six points are given, no three collinear. If all 15 line segments are drawn joining pairs of points, how many segments can be colored black without forming any black triangles?
6. You have a job promising a bar of gold for seven days' work. However, you may elect to stop working at the end of any one of the seven days. Can the bar be sliced at the beginning of the week with just two cuts to guarantee that you can be paid the appropriate amount owed you?

11 Episodes in the Calculation of Pi

Since prehistoric times, human beings have known well the shape we call a circle, based on the extreme importance of the sun and moon in everyday life. Many ancient cultures attempted to measure the circumference of a given circle and some standard values were long accepted. In this chapter we will get an overview of some of the efforts mathematicians from different epochs and various cultures have made at the somewhat elusive mensuration of the circle. If we consider a circle with unit diameter, we now denote its circumference by the Greek letter π (pi). By 2000 B.C.E. several cultures we collectively call the Babylonians (living in what is present day Iraq and Syria), used 3 1/8 as the value of π. In the Old Testament I Kings vii. 23, it is stated, "Also he made a molten sea of ten cubits from brim to brim, round in compass, and five cubits the height thereof; and a line of thirty cubits did compass it round about." The above passage refers to a vessel to be constructed by the bronze worker, Hiram of Tyre, who was hired by King Solomon. Here π is given as $\frac{30}{10} = 3$. Furthermore, there is little indication in either the Babylonian writings or in the Bible that the values of π are understood to be just approximations.

The realization that the area of a circle of unit radius is also π seems to be nearly as ancient. The Egyptians had a highly developed knowledge of geometry, which no doubt proved useful in their monumental pyramid constructions. Again there seems to be little or no distinction made between approximations and exact values. In the Rhind papyrus dated circa 1850 B.C.E., the scribe Ahmes stated that the area of a circle of diameter d is the same as the area of a square with sides of length $\frac{8}{9}d$. Since the area of the square is $\frac{64}{81}d^2$ and the area of the circle is $\pi(\frac{d}{2})^2$, it follows that the value for π was taken to be $\frac{256}{81}$. For practical usage, this is not a bad approximation. Furthermore, it led to one of the famous Greek problems of antiquity.

The three famous problems of antiquity were the following: 1) squaring the circle, 2) duplicating the cube, and 3) trisecting an angle. In all three the challenge is to construct a specified geometric object using only Euclidean tools, namely an unmarked straightedge and a collapsible compass. Squaring the circle involves

Mathematical Journeys, by Peter D. Schumer
ISBN 0-471-22066-3

the construction of a square of area equal to a given circle. Duplicating the cube requires the construction of a cube of volume precisely twice that of a given cube. The last problem demands a method to exactly trisect any given angle. These were three significant challenges that tested the resources and ingenuity of countless mathematicians from many lands over the course of the centuries. In fact, definitive solutions were not given until the 19th century! Furthermore, all answers turned out to be negative ones. That is, none of the three problems can be solved in general using just Euclidean tools. Nonetheless, many clever solutions have been found by other means with additional resources. But let's not get too far ahead of ourselves.

The ancient Greeks formalized much of geometry and further developed mathematics as a deductive science based on definitions, axioms, rules of logic, theorems, and explicit and verifiable proofs. The greatest of all the Greek geometers was Archimedes of Syracuse (287–212 B.C.E.). Archimedes made significant advances to all areas of mathematics and most areas of pure and applied science known in his time. In fact, the fields of hydrostatics and mechanics were essentially created by him. Most pertinent here is Archimedes's work, "Measurement of the Circle." In it the following proposition is rigorously proved: "The area of any circle is equal to that of a right triangle in which one of the sides equals the radius and the other side equals the circumference of the circle." What a beautiful theorem! Archimedes then proceeds to find a good approximation for the value of π. His method is intricate. Given a unit circle, he begins by inscribing and circumscribing it with regular hexagons and then calculates their perimeters. The inscribed hexagon has perimeter 6 and the circumscribed hexagon has perimeter $4\sqrt{3}$. It follows that $3 < \pi < 2\sqrt{3}$. Rather than work with square roots, Archimedes approximates them with fractions. In this case, Archimedes used the "fact" that $\frac{265}{153} < \sqrt{3} < \frac{1{,}351}{780}$. How he came up with this particular estimate has baffled mathematics historians ever since. (But it should be noted that both fractions appear as convergents in the infinite continued fraction for $\sqrt{3}$; and the notion of finite continued fractions is implicit in the Euclidean algorithm for finding greatest common divisors.) Next, Archimedes doubled the number of sides and got closer bounds for π by utilizing inscribed and circumscribed dodecagons (12-sided polygons). But Archimedes did not stop there. He continued his calculations with 24-sided polygons, then with 48-sided polygons, and finally with 96-sided polygons. For some reason, he felt satisfied stopping there. His final estimates squeezed π between the fractions $\frac{6{,}336}{2{,}017\frac{1}{4}}$ and $\frac{14{,}688}{4{,}673\frac{1}{2}}$. Since the first fraction is larger than $3\frac{10}{71}$ and the second fraction is smaller than $3\frac{1}{7}$, Archimedes summarized his result as

$$3\tfrac{10}{71} < \pi < 3\tfrac{1}{7}.$$

What a spectacular result! Even today the Archimedean value $3\frac{1}{7} = \frac{22}{7}$ is no doubt the most popular fractional approximation to π. Archimedes's value also gives us the decimal estimation of $\pi \approx 3.14$, accurate to two decimal places. In fact, in my experience a fair number of elementary school teachers seem to think that either $\frac{22}{7}$ or 3.14 is the actual value of π—or even worse, that both are exact values. Perhaps we aren't so different from the ancient Babylonians and Egyptians who failed to distinguish between exact and approximate formulas.

Archimedes

The idea of using multisided regular polygons to either inscribe or circumscribe a given circle was used by many subsequent mathematicians to make improvements on Archimedes's estimate. The Chinese scholar Liu Hui (ca. 260 C.E.) used a circle of radius 10 and inscribed polygons starting with a hexagon and working up to a 192-sided polygon. His work leads to $\pi \approx 3.1416$, accurate to four decimal places. A couple centuries later, the Chinese mathematician Tsu-ching Chih (ca. 480 C.E.) proceeded from where Liu Hui had left off, doubling the number of sides six more times. With a 12,288-sided polygon, he was able to establish that π lay between 3.1415926 and 3.1415927. He stated a slightly weaker, but visually striking result: $\pi \approx \frac{355}{113}$. This can be remembered by simply writing 113,355 and separating the number in the middle. For nearly 800 years this was the most accurate value known for π.

The polygon used by Tsu-Chih had $3 \cdot 2^{12}$ sides. In 1430, the Arab scholar Al-Kashi continued with calculations working up to a polygon with $3 \cdot 2^{28}$ sides and derived 16 accurate decimal places for π. The most impressive feat in this direction was undertaken by Ludolph van Cuelen, professor at the University of

Leyden. In 1596, he used a $60 \cdot 2^{33}$-sided polygon to surpass Al Kashi's record. But his obsession was not satisfied. After several years' progress, he announced a value of π accurate to 35 decimal digits based on a 2^{62}-sided polygon! When he died in 1610, his widow had the digits engraved on his tombstone.

The technique of using polygons with an ever-increasing number of sides is tiresome, arduous, and has its limitations. No matter how far we carry out this line of attack, we can only get an approximation to the value of π. Mathematicians began to wonder whether there was an exact formula for π as some infinite sum or product. Indeed there are many such results.

François Viète

The first such formula was due to the great French mathematician François Viète (1540–1603). Though by profession a lawyer and later member of parliament, Viète devoted his spare time to mathematics and had astonishing algebraic and analytical skills. He wrote on arithmetic, algebra, geometry, and trigonometry. He successfully met a challenge from the Dutch ambassador to France that required the solution of a 45th-degree polynomial. In addition, he deciphered a 400-character Spanish code which gave the French the decisive advantage in its war with Spain (1589–1590). King Phillip II of Spain allegedly complained to Pope Gregory XIV that the French were employing magic "contrary to the practice of the Christian faith." Viète was a tireless worker and for fun calculated π to ten decimal place accuracy by using polygons with $6 \cdot 2^{16} = 393{,}216$ sides. However, his main contribution in this direction was the first exact formula involving π, which we now present.

Theorem 11.1 (Viète, 1593): $\frac{2}{\pi} = \sqrt{\frac{1}{2}}\sqrt{\frac{1}{2} + \frac{1}{2}\sqrt{\frac{1}{2}}}\sqrt{\frac{1}{2} + \frac{1}{2}\sqrt{\frac{1}{2} + \frac{1}{2}\sqrt{\frac{1}{2}}}} \cdots$

Our proof of Theorem 11.1 utilizes a couple of more modern notions, but essentially all the basic ideas are those of Viète. One formula that we need is the double-angle formula from trigonometry, an area that Viète helped develop. The double-angle formula for sines states that

$$\text{For any angle } \theta, \sin 2\theta = 2\sin\theta\cos\theta. \tag{11.1}$$

Another formula that we'll require is the half-angle formula for cosines. In particular,

$$\text{For any angle } \theta, \cos\left(\frac{\theta}{2}\right) = \frac{1}{2}\sqrt{1 + \cos\theta}. \tag{11.2}$$

The final element that we will need is L'Hôpital's rule from calculus. It states that if f and g are differentiable functions in some neighborhood of a point c (including infinity) and if $\lim_{x\to c} f(x) = \lim_{x\to c} g(x) = 0$, then $\lim_{x\to c} \frac{f(x)}{g(x)} = \lim_{x\to c} \frac{f'(x)}{g'(x)}$ where f' and g' are the derivatives of f and g, respectively. Here I have skipped a couple of technical conditions so as not to get completely side-tracked.

Proof of Theorem 11.1: *Let $0 \le x \le \frac{\pi}{2}$. Letting $x = 2\theta$ in Equation 11.1 we get*

$$\sin x = 2\sin\left(\frac{x}{2}\right)\cos\left(\frac{x}{2}\right).$$

But we can apply Equation 11.1 to $\sin(\frac{x}{2})$ obtaining $\sin(\frac{x}{2}) = 2\sin(\frac{x}{4})\cos(\frac{x}{4})$. Repeated application of Equation 11.1 leads to

$$\sin x = 2^n \sin\left(\frac{x}{2^n}\right)\cos\left(\frac{x}{2}\right)\cos\left(\frac{x}{4}\right)\cdots\cos\left(\frac{x}{2^n}\right) \text{ for all } n \ge 1 \tag{11.3}$$

Next we evaluate $\lim_{n\to\infty} 2^n \sin(\frac{x}{2^n})$. To evaluate this limit, let $r = 2^n$ and note that as n approaches infinity, so does r. Hence

$$\lim_{n\to\infty} 2^n \sin\left(\frac{x}{2^n}\right) = \lim_{r\to\infty} r\sin\left(\frac{x}{r}\right).$$

Now let $t = \frac{1}{r}$. Then as r approaches infinity, t approaches zero (from the right). Thus

$$\lim_{r\to\infty} r\sin\left(\frac{x}{r}\right) = \lim_{t\to 0^+} \frac{\sin(xt)}{t}.$$

This last limit is of the indefinite form $\frac{0}{0}$. So we can apply L'Hôpital's rule to it. Hence

$$\lim_{n\to\infty} 2^n \sin\left(\frac{x}{2^n}\right) = \lim_{t\to 0^+} \frac{\sin(xt)}{t} = \lim_{t\to 0^+} \frac{x\cos(xt)}{1}$$

$$= x \text{ (since the derivative of sine is cosine).}$$

Next we repeatedly apply Equation 11.2:

$$\cos\left(\frac{x}{2}\right) = \sqrt{\frac{1}{2}(1+\cos x)}$$

$$\cos\left(\frac{x}{4}\right) = \sqrt{\frac{1}{2}\left(1+\cos\frac{x}{2}\right)}$$

$$= \sqrt{\frac{1}{2}\left(1+\sqrt{\frac{1}{2}(1+\cos x)}\right)}$$

$$\cos\left(\frac{x}{8}\right) = \sqrt{\frac{1}{2}\left(1+\cos\frac{x}{4}\right)}$$

$$= \sqrt{\frac{1}{2}\left(1+\sqrt{\frac{1}{2}\left(1+\sqrt{\frac{1}{2}(1+\cos x)}\right)}\right)}, \textit{ etc.}$$

Now let $x = \frac{\pi}{2}$ in Equation 11.3. Then

$$1 = \sin\frac{\pi}{2} = \frac{\pi}{2}\cdot\sqrt{\frac{1}{2}}\sqrt{\frac{1}{2}+\frac{1}{2}\sqrt{\frac{1}{2}}}\cdots$$

Multiplying both sides by $\frac{2}{\pi}$,

$$\frac{2}{\pi} = \sqrt{\frac{1}{2}}\sqrt{\frac{1}{2}+\frac{1}{2}\sqrt{\frac{1}{2}}}\sqrt{\frac{1}{2}+\frac{1}{2}\sqrt{\frac{1}{2}+\frac{1}{2}\sqrt{\frac{1}{2}}}}\cdots \qquad \square$$

Although Viète's result is not especially helpful in computing a good decimal value of π, it did usher in a new era for the discovery of beautiful formulas involving π. In England, John Wallis (1616–1703), a charter member of the Royal Society and holder of the prestigious Savilian chair of geometry at Oxford University, developed several trigonometric formulas that would later become a standard part of the corpus of integral calculus. In particular, in his *Arithmetica Infinitorum* (1655), reduction formulas for the integrals of arbitrarily high integer powers of sine are derived. From these, Wallis derived the following incredible result:

$$\frac{2}{\pi} = \frac{1}{2}\cdot\left(\frac{3}{2}\cdot\frac{3}{4}\right)\cdot\left(\frac{5}{4}\cdot\frac{5}{6}\right)\cdot\left(\frac{7}{6}\cdot\frac{7}{8}\right)\left(\frac{9}{8}\cdot\frac{9}{10}\right)\cdots\cdot \qquad (11.4)$$

The first president of the Royal Society, William Brouncker (1620–1684), manipulated Wallis's result to obtain a rather startling infinite continued fraction involving π. The regularity and clear sense of pattern in formulas 11.4 and 11.5 fueled the debate of whether π was an algebraic quantity (a root of some polynomial) or not (a transcendental number). Here is Brouncker's formula:

$$\frac{4}{x} = 1 + \cfrac{1^2}{2 + \cfrac{3^2}{2+\cfrac{5^2}{2+\cfrac{7^2}{2+\cdots}}}} \tag{11.5}$$

Another noteworthy contribution is due to Gottfried Wilhelm Leibniz (1646–1716), a universal genius and co-founder (along with Newton) of the calculus. Leibniz's result is more of an observation based on other known work. From the geometric series

$$\frac{1}{1-s} = 1 + s + s^2 + s^3 + \cdots$$

valid for $|s| < 1$, substitute $-s^2$ for s, obtaining

$$\frac{1}{1+s^2} = 1 - s^2 + s^4 - s^6 + \cdots .$$

Next integrate termwise,

$$\int_0^x \frac{1}{1+s^2} ds = x - \frac{x^3}{3} + \frac{x^5}{5} - \frac{x^7}{7} + \cdots .$$

Since the derivative of the arctangent function was known to be $\frac{1}{1+x^2}$, it follows that

$$\arctan x = x - \frac{x^3}{3} + \frac{x^5}{5} - \frac{x^7}{7} + \cdots .$$

This is known as Gregory's series, which appears in the work *Geometriae pars universalis* (1668) authored by the Scotsman James Gregory (1638–1675). Knowing that $\tan \frac{\pi}{4} = 1$, all Leibniz had to do was let $x = 1$ and argue why the resulting alternating series converges. Hence the Leibniz series becomes

$$\frac{\pi}{4} = 1 - \frac{1}{3} + \frac{1}{5} - \frac{1}{7} + \frac{1}{9} - \cdots \tag{11.6}$$

Again the wonderful series above is of little practical use, given the slow convergence of the series, but as a work of art it's a masterpiece!

The beginning of the next era for π computations can be dated to 1706, the year that John Machin (1680–1751) published his ground-breaking research on calculating π. (By the way, Machin is pronounced mā-chan.) Probably aware of

Leibniz's work, Machin studied the arctangent function more carefully. Here's what he did:

$$\text{Let } \alpha = \arctan\left(\frac{1}{5}\right) = \frac{1}{5} - \frac{1}{3\cdot 5^3} + \frac{1}{5\cdot 5^5} - \frac{1}{7\cdot 5^7} + \cdots .$$

By the angle addition formula for tangent,

$$\tan 2\alpha = \frac{2\tan\alpha}{1-\tan^2\alpha} = \frac{2/5}{1-1/25} = \frac{5}{12}.$$

Similarly,

$$\tan 4\alpha = \frac{2\tan 2\alpha}{1-\tan^2 2\alpha} = \frac{5/6}{1-25/144} = \frac{120}{119}.$$

So 4α is approximately $\frac{\pi}{4}$ since $\tan\frac{\pi}{4} = 1$. But Machin didn't stop here. Let $\beta = 4\alpha - \frac{\pi}{4}$, the slight angle difference between 4α and $\frac{\pi}{4}$. Applying the angle addition formula for tangent again,

$$\tan\beta = \frac{\tan 4\alpha - \tan\dfrac{\pi}{4}}{1+\tan 4\alpha\cdot\tan\dfrac{\pi}{4}} = \frac{\dfrac{120}{119}-1}{1+\dfrac{120}{119}\cdot 1} = \frac{1}{239},$$

$$\text{so } \beta = \arctan\frac{1}{239} = \frac{1}{239} - \frac{1}{3\cdot 239^3} + \frac{1}{5\cdot 239^5} - \cdots .$$

Since $\frac{\pi}{4} = 4\alpha - \beta$, Machin's series becomes

$$\frac{\pi}{4} = 4\left[\frac{1}{5} - \frac{1}{3\cdot 5^3} + \frac{1}{5\cdot 5^5} - \frac{1}{7\cdot 5^7} + \cdots\right] - \left[\frac{1}{239} - \frac{1}{3\cdot 239^3} + \frac{1}{5\cdot 239^5} - \cdots\right] \tag{11.7}$$

Machin's formula is not only attractive, it is also highly practical for computing π. The arithmetic required to divide by ascending powers of 5 is easily handled due to the simple terminating decimal expansions of such fractions. The decimal expansions of the terms beginning with $\frac{1}{239}$ are much more cumbersome, but the rapid growth of their denominators means that fewer of their terms must be taken into account in any particular computation. Machin now had a practical formula that could compete with and defeat all previous methods dependent on multisided polygons. In fact, after significant effort, Machin calculated π to 100 decimal places. At this point, he probably thought that would be the end of the story. Why would anyone endeavor to go further? Well, human nature being what it is ....

Other mathematicians soon derived other arctangent formulas, several of which had practical application to the computation of π. One fine example is due to the prolific Leonhard Euler, who played a prominent role in so many areas of mathematics (as evidenced throughout this book). Euler developed a family of series for the arctangent function, namely

$$\arctan x = \frac{y}{x}\left[1 + \frac{2}{3}y + \frac{2 \cdot 4}{3 \cdot 5}y^2 + \frac{2 \cdot 4 \cdot 6}{3 \cdot 5 \cdot 7}y^3 + \cdots\right]$$

where the dependent variable $y = \frac{x^2}{1+x^2}$.

If $x = \frac{1}{7}$, then $y = \frac{1}{50} = 0.02$, and calculations are readily done. Better yet, let $x = \frac{3}{79}$ so that $y = \frac{144}{10{,}000} = 0.0144$. With this value, Euler calculated 20 decimal places of π in less than half an hour—no doubt just a little diversion before getting down to some real work!

During the 18th and 19th centuries, a host of new arctangent formulas were developed with hopes of extending the decimal expansion of π. One noteworthy success was made in 1844 by the phenomenal lightning calculator, Zachariah Dase (1820–1861), who computed π to 200 decimal places. Dase was a human calculator and had been hired by the Hamburg Academy of Sciences to extend current logarithm tables. In addition, Gauss himself had suggested that Dase might be useful in extending the table of known primes, a task he accepted and successfully completed for primes up to 9,000,000. For the π calculation, Dase used a new arctangent formula due to the Austrian Lutz von Strassnitzky (1803–1852), who derived this nice formula:

$$\frac{\pi}{4} = \arctan\frac{1}{2} + \arctan\frac{1}{5} + \arctan\frac{1}{8}$$

By this time, much more was known about the arithmetic nature of the number π. In 1761, Johann Lambert (1728–1777) proved that both π and π^2 were irrational. Lambert was a man of enormous talents who made contributions to astronomy, cartography, the nature of heat, philosophy, acoustics, and several areas of mathematics in over 150 published scholarly papers. He even named the hyperbolic sine and cosine functions. His contribution to π showed that no rational fraction would ever equal π exactly. But the ancient question of whether or not a circle could be squared with Euclidean tools still remained open. That question was disposed of negatively in 1882 by the German Ferdinand Lindemann (1852–1939) when he demonstrated that π is transcendental. Hence no polynomial with rational coefficients has π as a root. A more modern open question asks whether π is *normal*. In other words, do all sequences of decimal digits appear in the expansion of π to the degree statistically expected? If so, then π is a normal number. For example, does the digit 7 ultimately appear one-tenth of the time in the decimal expansion of π? Interestingly, this reasonable question is still an open one.

Modern digit hunters of π use very sophisticated mathematics that generally was not available to earlier mathematicians. In addition, the awesome speed and power of present-day computers (sometimes working together simultaneously) must be employed. But there is an exception to the notion that the right

Carl Friedrich Gauss

mathematics didn't exist earlier. The work of Carl Friedrich Gauss (1777–1855) on the arithmetic-geometric mean inequality has proven extremely amenable to modern computational methods. In the years 1791–1792, while investigating the arclength of the lemniscate, Gauss began to develop an algorithm which in theory would give *quadratic* convergence to π, that is, the number of accurate decimal places doubles after each iteration. Here are some details.

Let $a_0 = a$ and $b_0 = b$ be our initial values. Furthermore, for $k \geq 1$, let $a_{k+1} = \frac{a_k + b_k}{2}$ be the arithmetic mean of the previous values and let $b_{k+1} = \sqrt{a_k b_k}$ be their geometric mean. Then both a_k and b_k converge very rapidly to a number denoted AGM (a, b), where AGM stands for "arithmetic-geometric mean." By letting $a = \sqrt{2}$ and $b = 1$, Gauss calculated a_4 and b_4 obtaining the following values: $a_4 = 1.19814023473559220744 1\ldots$ and $b_4 = 1.198140234735592207439\ldots$. Notice the amazing agreement up to the 20th decimal place! Later Gauss was able to show that

$$\frac{\pi}{2\int_0^1 dz/\sqrt{1-z^4}} = \text{AGM}\,(1, \sqrt{2}) \tag{11.8}$$

Though Gauss did not have the computer resources we now enjoy, his work provided the theoretical key that would later allow researchers to calculate hundreds of millions—even billions—of digits of π. In fact, the current world record is held by two Japanese computer scientists, Yasumasa Kanada of the University of Tokyo together with assistance from Daisuke Takahashi. In October of 1999, using a version of Gauss's AGM formula, they calculated 206,158,430,000 digits of π! The computation took about 80 hours on a Hitachi SR 8000 supercomputer. But more impressive is that it relied on some nearly forgotten pure mathematics of a brilliant German mathematician from 200 years earlier. Great ideas do not go out of date.

Srinivasa Ramanujan

Another example of a mathematician who was frighteningly ahead of his time is Srinivasa Ramanujan (1887–1920), an Indian mathematician of incomparable perspicacity. In fact, modern-day mathematicians are still unraveling the deep and sometimes enigmatic mathematics left in his notebooks. Ramanujan grew up in poverty in southern India, had precious few mathematical resources (books or teachers), and little encouragement from family or friends. Yet with astonishing determination and single-mindedness, his genius blossomed, trampled all impediments in its way, and propelled Ramanujan to become one of the foremost creative mathematicians of the early 20th century. Much of his best work was done at Cambridge University between 1913 and 1919, a substantial portion together with G.H. Hardy. Ramanujan's work includes significant discoveries in analytic number theory, the theory of primes, the partition function, the representation of numbers as sums of squares, new definite integral formulas, and error bounds in lattice point problems. He also made highly original contributions to the esoteric area of elliptic and modular forms and to a new class of functions he called mock theta functions. Here I simply report two of his results that related directly to the number π.

One such startling result is the following:

$$\frac{1}{\pi} = \sum_{n=0}^{\infty} \frac{(2n)!^3}{n!^6} \frac{4^{2n+5}}{2^{12n+4}}.$$

In a 1914 paper, "Modular Equations and Approximations to π," Ramanujan gave 30 different formulas for π, including this one, which gives about eight new decimal places for each term of the series:

$$\frac{1}{\pi} = \frac{\sqrt{8}}{9801} \sum_{n=0}^{\infty} \frac{(4n)!}{n!^4} \frac{1{,}103 + 26{,}390n}{396^{4n}}.$$

Such formulas have inspired the most talented modern-day mathematicians, who have built upon and extended Ramanujan's work. Among the most ardent of π digit hunters are the brothers David and Gregory Chudnovsky, Russian immigrants in the United States. Here is one of their Ramanujan-like formulas, which gives 15 decimal places of π per term:

$$\frac{1}{\pi} = \frac{12}{\sqrt{640{,}320^3}} \sum_{k=0}^{\infty} (-1)^k \frac{(6k)!}{k!^3(3k)!} \frac{13{,}591{,}409 + 545{,}140{,}134k}{(640{,}320)^{3k}}.$$

At one point the Chudnovskys had turned Gregory's apartment into a veritable supercomputer built by the brothers themselves from parts they purchased at a local Radio Shack and dedicated solely to cranking out ever more digits of π. In fact, the Chudnovskys have been in a back and forth race with Yasumasa Kanada since the early 1980s. The Chudnovskys were the first to break the billion mark for π digits, and have held the world record at least six times. Though it's a bit difficult to really keep track since both the Chudnovskys and Kanada are rather private about their techniques and latest pursuits.

Another pair of brothers who have made enormous strides in our understanding of ways to compute π are the Canadians Jonathan and Peter Borwein, both currently at Simon Fraser University in British Columbia. In 1994 they published a complicated Ramanujan-type formula that gave 50 digits per term. Of even greater utility to π digit hunters was an iterative algorithm they had developed in 1987 known as Borweins's Quartic Formula. Specifically, let $y_0 = \sqrt{2} - 1$ and $a_0 = 6 - 4\sqrt{2}$ be the initial values. For $k \geq 0$, let

$$y_{k+1} = \frac{1 - \sqrt[4]{1 - y_k^4}}{1 + \sqrt[4]{1 - y_k^4}} \text{ and } a_{k+1} = a_k(1 + y_{k+1})^4 - 2^{2k+3} y_{k+1}(1 + y_{k+1} + y_{k+1}^2).$$

Then a_{k+1} converges very rapidly to $\frac{1}{\pi}$. In fact, each iteration quadruples the number of accurate digits, hence the moniker "quartic" formula. This was the specific formula that Kanada used in several of his record-breaking computations of π, including a 68 billion digit computation in early 1999. It's interesting that just a few months later he switched back to Gauss's AGM formula for the current record.

Our last episode in π calculation involves new formulas that allow individual bits of π to be computed without having to find all the preceding bits. By bit we are referring to binary digits. After all, for a computer, base two is more natural than base ten. The startling discovery that such computations were even possible was due to David Bailey, Peter Borwein, and Simon Ploufe—all of Simon Fraser University. In 1995 they announced the following formula for π, now called the BBP algorithm:

$$\pi = \sum_{n=0}^{\infty} \frac{1}{16^n} \left(\frac{4}{8n+1} - \frac{2}{8n+4} - \frac{1}{8n+5} - \frac{1}{8n+6} \right).$$

The occurrence of 16^n in the denominator of every term allows for the remarkable ability to jump to any hexadecimal (base 16) digit of π. But each hexadecimal digit can be expanded to four bits since $2^4 = 16$, and so four times as many bits of π can be so obtained. Even though no such formulas involving powers of ten have been discovered, similar identities to BBP have been developed with the aid of computer algebra systems. One especially aesthetically appealing one is due to Victor Adamchik and Stan Wagon (1997):

$$\pi = \sum_{n=0}^{\infty} \frac{(-1)^n}{4^n} \left(\frac{2}{4n+1} + \frac{2}{4n+2} + \frac{1}{4n+3} \right).$$

How far out can we go and grab a single bit of π? The formula that's been used the most often is the following one involving powers of $1{,}024 = 2^{10}$ due to the French mathematician Fabrice Bellard:

$$\pi = \frac{1}{64} \sum_{n=0}^{\infty} \frac{(-1)^n}{1{,}024^n} \left(\frac{-32}{4n+1} - \frac{1}{4n+3} + \frac{256}{10n+1} - \frac{64}{10n+3} - \frac{4}{10n+5} - \frac{4}{10n+7} + \frac{1}{10n+9} \right).$$

A graduate student at Simon Fraser University (now studying at Oxford University) holds the current world record. Colin Percival learned all his high school mathematics before completing seventh grade. That year he took the 12th grade "Euclid Contest" and got the highest score in British Columbia. By tenth grade, Percival was competing in the Putnam Examination. His score that year placed him among the top 60 college finishers in all of United States and Canada. The next year he ranked 12th and his last year of high school he became a Putnam Fellow, scoring one of the top six scores in North America.

Percival's audacious project involved cleverly pooling the resources of thousands of computer users over the Internet. By farming out the calculations and sharing the capabilities of 1,734 computers from 56 different nations, Percival was able to productively use 1.2 million CPU hours of what would otherwise be computer idle time. The project itself lasted from September 5, 1998, until September 11, 2000. On that day, Percival was able to complete the calculation of the 250 trillionth hexadecimal digit of π. By converting it into binary, the quadrillionth bit of π was thus determined! It's a zero. Some may quip that that's a lot of work, all for nothing. But I find it highly appropriate that the long search to understand the circle has ended up with the perfect digit, namely the one we denote with just such a round circle.

WORTH CONSIDERING

1. Convert to decimal expansions the ancient approximations of π: Babylonia–$3\frac{1}{8}$, Egypt–$\frac{256}{81}$, China–$\sqrt{10}$, India–$4 \cdot (\frac{9{,}785}{11{,}136})^2$, Ptolemy–$\frac{377}{120}$, Archimedes–$\frac{22}{7}$, Tsu-ching Chih–$\frac{355}{113}$.

2. Plato is credited with discovering the approximation $\sqrt{2}+\sqrt{3}$ for π. Derive this result by averaging the perimeters of an inscribed square and a circumscribed hexagon of a given circle. What decimal place accuracy do we get?

3. (Due to Franz Gnädinger)

(a) Approximate π by taking a circle of radius 5 centered at the origin and inscribing it with an irregular dodecagon (12-sided polygon) with vertices at (0, ±5), (±5, 0), (±3, ±4), and (±4, ±3).

(b) Expand the circle in part (a) five-fold to obtain one of radius 25 centered at the origin. The points corresponding to (0, 5), (3, 4), (4, 3), and (5, 0) now have coordinates (0, 25), (15, 20), (20, 15), and (25, 0). In the first quadrant add in the points (7, 24) and (24, 7) and do the same in the other three quadrants. Now use this irregular inscribed icosagon (20-sided polygon) to estimate π.

(c) Expand the circle again five-fold. In addition to the expanded vertices, add in the points (±44, ±117) and (±117, ±44). Approximate π once again. This clever method, based on Pythagorean triplets, can be continued indefinitely. Note that all calculations depend only on knowing the expansion of $\sqrt{2}$, $\sqrt{5}$, and their product.

4. Use the angle addition formulas for sine and cosine to derive

(a) $\sin x = 2\sin(\frac{x}{2})\cos(\frac{x}{2})$ and

(b) $\cos\frac{x}{2} = \sqrt{\frac{1}{2}(1+\cos x)}$, two formulas required in Viète's derivation.

5. Verify Leonardo da Vinci's (1452–1519) construction: Roll a wheel of radius r and tire width $\frac{r}{2}$ one revolution over an impressionable surface. The resulting rectangle has area πr^2.

6. **(a)** Use the angle addition formula for tangent, namely $\tan(a+b) = \frac{\tan a + \tan b}{1 - \tan a \cdot \tan b}$, to verify the following arctangent formula:

$$\arctan\frac{1}{a-b} = \arctan\frac{1}{a} + \arctan\left(\frac{b}{a^2 - ab + 1}\right).$$

(b) Use the formula above with $a = 2$ and $b = 1$ to derive an arctangent formula of Euler's (1738).

(c) Use (b) plus the formula above with $a = 3$ and $b = 1$ to derive an arctangent formula due to Charles Hutton (1776).

7. Use Gauss's AGM formula to calculate a_k and b_k for $k = 1$, 2, and 3 with $a_0 = 1$ and $b_0 = \sqrt{2}$.

8. Determine the number of accurate decimal places obtained from the following estimates of π.

(a) (Ramanujan) $\frac{355}{113}\left(1 - \frac{3}{35{,}330{,}000}\right)$

(b) $\frac{47^3+20^3}{30^3} - 1$.

9. (R.G. Duggleby) Comment on the value of $\sqrt[6]{\pi^4 + \pi^5}$.

10. (Daniel Shanks)

(a) Show that if p_k is an approximation to π accurate to n decimal places, then $p_{k+1} = p_k + \sin p_k$ is accurate to at least $3n$ decimal places.

(b) Use the above method with $p_0 = 3$ to obtain at least nine decimal place accuracy for π.

11. Continued Fraction for π: Let $a_0 = 3$, $a_1 = 7$, $a_2 = 15$, $a_3 = 1$, $a_4 = 292$, $a_5 = 1$, $a_6 = 1$, $a_7 = 1$, $a_8 = 2$, $a_9 = 1$, and $a_{10} = 3$. Define the *convergents* to π by $c_0 = a_0$, $c_1 = [a_0; a_1] = a_0 + \frac{1}{a_1}, \ldots, c_k = [a_0; a_1, \ldots, a_{k-1} + \frac{1}{a_k}]$ for $k \geq 2$. Calculate the first ten convergents to π.

12 A Sextet of Scintillating Problems

There are few pleasures as delicious and yet enduring as a really good mathematical problem. In this chapter, I present six interesting problems from different areas of mathematics. None of them are trivial; yet luckily, none are at all inaccessible. Better yet, each can serve as an entrée to other related problems and new, significant areas of mathematics. Try to solve them yourself first, then carefully read the solutions. I think you will agree that each one is both fun and instructive. At the International Congress of Mathematicians in 1900, David Hilbert said, "A mathematical problem should be difficult to entice us, yet not be completely inaccessible, lest it mock at our efforts. It should be a guide post on the mazy path to hidden truths, and ultimately a reminder of our pleasure in the successful solution." Hopefully, these problems will share some of these lofty attributes.

Problem #1: *Given a positive integer m, does there exist a circle in the plane having exactly m interior lattice points?*

Recall that the lattice points are the points (a, b) where both a and b are integers. This question was asked by the outstanding Polish mathematician Hugo Steinhaus (1887–1972) in 1957. Steinhaus was born in Jaslo, Galicia, to a Jewish intellectual family. He did his graduate work at the University of Göttingen in Germany under the direction of Hilbert. Steinhaus did highly original work in functional analysis and was especially interested in its applications to orthogonal series and to probability. He helped develop measure theory and wrote the first papers on game theory as well. But his greatest contributions might be his influence on fellow mathematicians and his love of sharing mathematical insights with friends and colleagues. In 1916 he formed the Polish Mathematical Society in Kraków with several like-minded young mathematicians. From a humble beginning of weekly group discussions in a local park, their collaborations eventually led to the development of new areas of analysis and a great number of publications. Even more famous were the mathematical meetings years later at the Szkocka Café, commonly known as the Scottish Cafe, in Lvov (then Poland, now

Mathematical Journeys, by Peter D. Schumer
ISBN 0-471-22066-3

Ukraine). Beginning in 1935 and extending well into 1941, Steinhaus together with S. Banach, S. Ulam, S. Mazur, M. Kac, and several other mathematicians met and discussed significant mathematical problems several evenings a week at this cafe. Once solved, the problems were written up in a notebook that was kept there by a headwaiter at the cafe. When the Russians invaded, the meetings continued and some contributions to the *Scottish book* were even made by several visiting Russian mathematicians. To a get a taste of one of the problems, a famous one due to Steinhaus is the fair apportionment problem. If n people want to share a cake, how can they cut it so that all persons are satisfied that they have a fair-sized piece. Steinhaus provided a solution for $n = 3$, but a general solution was not worked out until the 1990s. The last entry, number 193, was made on May 31, 1941, just before the German invasion. Despite the fact that more than half of the Polish university mathematicians died or were killed during the next four years, the Scottish book survived the war and has since been translated into several languages. Fortunately for humanity's sake, great ideas don't die so easily.

Problem #1 is not part of the Scottish book, but is rather a later problem. It was solved almost immediately by Waclaw Sierpinski (1887–1969), a noted Polish mathematician and no slouch himself. Sierpinski did significant work in real analysis, number theory, topology, and set theory. His first paper, completed in 1904, dealt with Gauss's famous Circle Problem. Let $R(r)$ be the number of lattice points inside the circle centered at the origin with radius r. In 1837, Gauss had proved that $|R(r) - \pi r^2| < Cr$ for some constant C. Sierpinski made the first significant improvement to this inequality by proving that $|R(r) - \pi r^2| < Cr^{2/3}$. Since then, many new techniques have been developed leading to further improvements to the exponent in the upper bound. However, it should be noted that there is a limit due to a result of Hardy and Landau that says $|R(r) - \pi r^2| > Cr^{1/2}$. In any event, Sierpinski's theorem resulted in his receiving the Gold Prize that year from the University of Warsaw together with a college degree. Over the course of an illustrious career, Sierpinski published 724 papers, 50 books, and managed a great deal of editorial work for several eminent journals, including *Acta Arithmetica*, which he founded in 1958.

Solution to Problem #1: *The answer to the question is "Yes." In fact, we will show that the particular point $p = (\sqrt{2}, \frac{1}{3})$ has different distances from all lattice points. Thus we can choose an appropriate radius so that such a circle centered at p will contain exactly m lattice points for any given m. In particular, if the set of lattice points $\mathbf{Z}^2 = \{p_1, p_2, \ldots\}$ is ordered by increasing distance from p, then the circle consisting of the set of all x for which $|p - x| < |p - p_{m+1}|$ contains exactly the lattice points $p_1, \ldots, p_m$. Here $|x - y|$ means the distance between x and y. Figure 12.1 illustrates this for $m = 3$.*

To prove that the lattice points can be so ordered, for the sake of argument let us assume otherwise. Then there would be two distinct lattice points $a = (a_1, a_2)$ and $b = (b_1, b_2)$ such that $|p - a| = |p - b|$ with a_1, a_2, b_1, and b_2 all integers.

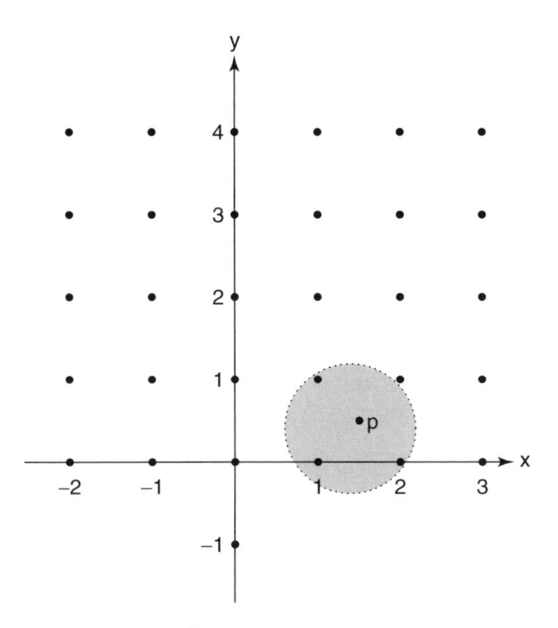

Figure 12.1 Circle centered at ($\sqrt{2}$, 1/3) containing exactly three lattice points.

But then

$$(a_1 - \sqrt{2})^2 + (a_2 - \tfrac{1}{3})^2 = (b_1 - \sqrt{2})^2 + (b_2 - \tfrac{1}{3})^2.$$

This implies that

$$a_1^2 + a_2^2 - b_1^2 - b_2^2 - \tfrac{2}{3}a_2 + \tfrac{2}{3}b_2 = 2(a_1 - b_1)\sqrt{2}. \quad (12.1)$$

But the number $\sqrt{2}$ is irrational, and hence so is the right-hand side of Equation 12.1 as long as $2(a_1 - b_1)$ is nonzero. But the left-hand side is rational. Hence the only way to rectify this is if $a_1 - b_1 = 0$, that is, $a_1 = b_1$. Hence, the left-hand side of Equation 12.1 is zero as well. Thus

$$a_2^2 - b_2^2 - \tfrac{2}{3}(a_2 - b_2) = 0. \quad (12.2)$$

If $a_2 = b_2$, then the points a and b are not distinct. Hence assume $a_2 \neq b_2$. But Equation 12.2 can be rewritten as

$$(a_2 - b_2)(a_2 + b_2 - \tfrac{2}{3}) = 0.$$

It follows that $a_2 + b_2 - \frac{2}{3} = 0$, or that $a_2 + b_2 = \frac{2}{3}$. But this contradicts the fact that a_2 and b_2 are integers. The result follows. □

Isn't it beautiful that the solution is so concrete and constructive. Furthermore, the only preliminary mathematics needed was the result, well-known by the ancient Greeks, that $\sqrt{2}$ is irrational. Several related results were established shortly after Sierpinski proved this one. Steinhaus extended the result by showing that for each m there is a circle of area precisely m containing exactly m lattice points. Then A. Schinzel proved that for all m there is a circle C_m with m lattice points on its circumference. Next T. Kulikowski established an analogous result on the surface of a sphere. Such is the way of mathematics and mathematicians. You might find it of interest to investigate what other values of p work in Sierpinski's proof. In addition, instead of circles, what about squares or other polygonal shapes or cubes in three dimensions? Generalizing the result from lattice points to "rational points," where both coordinates are rational numbers, leads in all sorts of new directions and can serve as an introduction to the wonderful world of algebraic geometry. Needless to say, Problem #1 can be one step toward an endless journey.

Problem #2: *Can the natural numbers be partitioned into two subsets so that each contains arithmetic progressions of every finite length, but neither contains an arithmetic progression of infinite length?*

Play around with this one for a bit before proceeding. You may find a solution.

Solution to Problem #2: *List all the natural numbers consecutively in a triangle as shown in Figure 12.2. Partition the positive integers into two sets, A and B, as follows. Counting the top row as row number one, let A consist of all integers in the odd-numbered rows and let B consist of all the integers in the even-numbered rows. Hence* $A = \{1, 4, 5, 6, 11, 12, 13, 14, 15, 22, 23, \dots\}$ *and* $B = \{2, 3, 7, 8, 9, 10, 16, 17, 18, \dots\}$. *Clearly each has an arithmetic progression (of difference 1) of length n for each n. However, no arithmetic progression in either A or B goes on indefinitely. The reason is that for any arithmetic progression of difference d, say, eventually there will be a gap exceeding d and gaps continue to get larger. So all arithmetic progressions will reach a void they can't jump across.* □

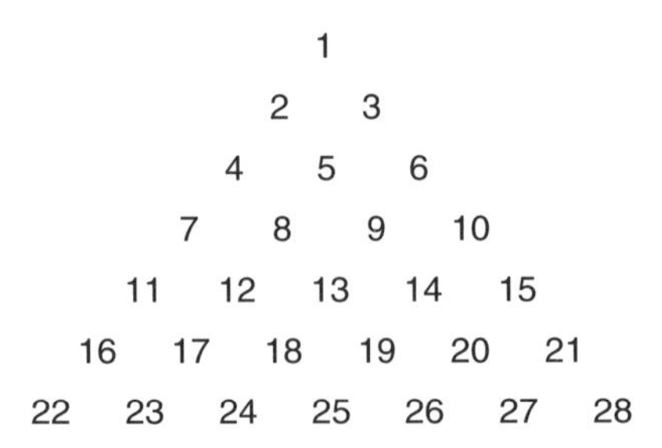

Figure 12.2 Natural numbers listed in triangular fashion.

Problem #2 provides a good example of the distinction between "arbitrarily large" versus "infinite." Although the arithmetic progressions can be as large or larger than any prescribed number, they never go on forever.

A deeper and more difficult problem was conjectured by the Dutch mathematician P.J.H. Baudet in 1926. He conjectured that if the positive integers are partitioned into *any* two subsets, then at least one of the subsets must contain an arithmetic progression of length l no matter how large l is. This problem immediately attracted the attention of several mathematics professors and students at the University of Göttingen. With some insights provided by E. Artin and O. Schreier, a young Dutch graduate student, B.L. van der Waerden (1903–1996) solved the problem later that same year, validating Baudet's conjecture. Since then an entire area of mathematics known as Ramsey theory has grown out of the study of similar problems. Van der Waerden taught widely in Germany, the United States, Switzerland, Holland, and elsewhere. He had over 40 doctoral students during his time in Zurich alone. Van der Waerden was a true polymath, publishing significant results in algebraic geometry, abstract algebra, topology, number theory, geometry, combinatorics, analysis, probability theory, mathematical statistics, and quantum mechanics. In addition, he wrote a great deal on the history of mathematics and astronomy and made important discoveries about the origins of algebra and geometry.

Problem #3: *What is the largest integer not expressible as the sum of five nonzero squares?*

Problem #3 might seem a bit odd without proper motivation, but it's a natural number theoretic question. If you pick a few natural numbers and write them as the sum of as few squares as possible, you'll soon notice that every one can be expressed as the sum of four or fewer squares. This observation is implicit in the work of Diophantus (ca. 250 C.E.) and that of many mathematicians from that time on. Interestingly, it wasn't until 1770 that anyone actually proved that every positive integer is expressible as the sum of at most four squares. Joseph Louis Lagrange (1736–1814) provided the first proof, followed by several others by Euler. The additional condition of expressing integers with nonzero squares changes the nature of the problem. Even though every positive integer can be expressed as the sum of four squares allowing zero as a perfect square, many integers cannot be expressed as the sum of three, four, five, or some other number of nonzero squares. In fact, there are infinitely many integers not expressible as the sum of k nonzero squares for any fixed $k \leq 4$. Note that the number 2 is not representable as the sum of four nonzero squares. Furthermore, if n is odd and $2n$ is not representable as the sum of four nonzero squares, then neither is the number $4^a \cdot 2n$ for all $a \geq 1$. (We leave this assertion as an interesting exercise.) Hence there are infinitely many such integers.

From the preceding discussion, it isn't *a priori* clear that there even is a largest integer not expressible as the sum of five nonzero squares. Conceivably the list of integers not so representable might be infinite, as was the case for four or fewer squares. However, we will see that this is not the case. Problem #3 and its solution is due to E. Dubouis (1911).

Solution to Problem #3: *The first few integers representable as the sum of five nonzero squares are 5, 8, 11, 13, 14, and so on. There are many gaps in our list, including the number 33, which cannot be written as the sum of five nonzero squares (which you should verify). After that, all numbers seem to have such a representation. But how do we prove that such a condition is satisfied forever?*

It is necessary to check all numbers up to and including 169 verifying that 33 is the last number not expressible in the desired manner. For any integer $n > 169$, we can write $n - 169$ as $a^2 + b^2 + c^2 + d^2$ for integers $a, b, c, d \geq 0$ by Lagrange's theorem. There are four cases to consider depending on the number of nonzero entries among a, b, c, and d.

(i) *If* $a, b, c, d \neq 0$, *then* $n = 13^2 + a^2 + b^2 + c^2 + d^2$.
(ii) *If only* $d = 0$, *then* $n = 12^2 + 5^2 + a^2 + b^2 + c^2$.
(iii) *If only* $c = 0$ *and* $d = 0$, *then* $n = 12^2 + 4^2 + 3^2 + a^2 + b^2$.
(iv) *If only* $b = 0, c = 0$, *and* $d = 0$, *then* $n = 10^2 + 8^2 + 2^2 + 1^2 + a^2$.

Therefore, 33 is the largest integer not expressible as the sum of five nonzero squares. □

As you can well imagine, Problem #3 is just the tip of the iceberg, and unlike the voyagers aboard the Titanic, I invite you to crash into it and safely explore. Here are some questions that come to mind: What is the smallest integer expressible in two ways as the sum of five nonzero squares? How about three ways or four ways, etc.? What, if any, is the largest integer not expressible in two (three, four, ...) ways as the sum of five nonzero squares? What about sums of six squares, or seven squares, etc.? It turns out that for all $k \geq 5$, all but a finite number of integers are sums of k nonzero squares. What's the situation with sums of distinct squares? Roland Sprague (1948) proved that every integer greater than 128 is the sum of distinct squares. And what about cubes, triangular numbers, primes, etc.? There's lots to investigate and the only tool you need is your imagination.

Problem #4: *Competing in the world professional go championship are 60 players from Tokyo, 20 from Beijing, 15 from Seoul, 10 from Osaka, 5 from Shanghai, 2 from Taipei, 1 from London, 1 from Paris, and 1 from San Francisco. Where should the championship be held to be the fairest from a collective players' prospective?*

We must first agree on what is meant by *fairest*. We could somehow average all the latitudes and longitudes of all the participants. Perhaps even take a weighted average since some cities have larger representation. We might instead pick the point on the earth that minimizes the maximum distance traveled by all the contestants. However, we choose an even simpler definition, and arguably the best. We will assume that the fairest location is the one that minimizes the total intercity travel of all the participants.

This problem is a classic math puzzle usually cloaked as a chess championship problem where most of the competitors come from New York City. Much to the consternation of the non-New Yorkers, it turns out that having the chess championship in New York is the fairest location. Similarly, in our version the answer is Tokyo. I enjoy playing go more than I do playing chess, so I've changed the problem to suit my interests. But other hobbies are fair game as well. Checkers anyone?

Solution to Problem #4: *Let $T_1, \ldots, T_{60}$ be the Tokyo players and let $P_1, \ldots, P_{55}$ be the others. Pair up the players $(T_1, P_1), (T_2, P_2), \ldots, (T_{55}, P_{55})$ leaving $T_{56}, \ldots, T_{60}$. (This pairing has nothing to do with match play itself.) Now let T_iP_i be the distance from Tokyo to player P_i's city for $1 \leq i \leq 55$. For each i, players T_i and P_i must together travel at least T_iP_i since that is the distance the two must travel at any point directly between or at either of their home cities. Any other location would require the pair of them to travel even further. So the total distance traveled by all the players is at least the sum $T_1P_1 + \ldots + T_{55}P_{55}$. If the site is Tokyo, then this is the exact total distance traveled since $T_{56}, \ldots, T_{60}$ do not have to travel at all. Anywhere else will increase the total distance. So Tokyo is the fairest site after all.* □

Problem #5: *A skyscraper has 101 floors numbered 1 to 101. Suppose that an elevator stops 51 times as it descends from the top floor. Show that it stops at two floors whose sum is 101.*

The solution to Problem #5 uses one of the simplest, but surprisingly useful, techniques in all of mathematics, namely the pigeonhole principle. The *pigeonhole principle* states that if there are n pigeonholes holding more than n pigeons, then there must be at least one pigeonhole with more than one pigeon. Even a bird brain could understand that. Equivalently, if n sets collectively contain more than n elements, then at least one of the sets contains two or more elements. And, oh yes, our building does have a floor number 13.

Solution to Problem #5: *Suppose that the elevator stops at floors numbered $f_1, \ldots, f_{51}$ where $1 \leq f_1 < f_2 < \ldots < f_{51} < 101$. Now consider the numbers $f_1, f_2, \ldots, f_{51}, 101 - f_1, 101 - f_2, \ldots, 101 - f_{51}$. These 102 numbers all lie between 1 and 100 inclusive. By the pigeonhole principle, they cannot all be distinct. But none of the f_i's are equal to each other, and so none of the $101 - f_i$'s are equal either. It necessarily follows that there is an i and j with $f_i = 101 - f_j$. Hence $f_i + f_j = 101$ as desired. Finally, note that f_i and f_j are in fact distinct, for otherwise they would both be equal to $50\frac{1}{2}$, not a place where anyone but a pigeon would want to get off the elevator.* □

Problem #6: *Call a positive integer square-full if each of its prime factors appears to at least the second power (e.g., $108 = 2^2 \cdot 3^3$ is square-full). Prove that there are infinitely many pairs of consecutive square-full numbers.*

This is a delightful problem with a surprisingly brief solution. So you might want to begin by seeing if you can list some square-full numbers and then look for at least one such consecutive pair. It shouldn't take too long. Hint: the smallest such pair are each one digit numbers!

Solution to Problem #6: *We prove the result inductively. Notice that $8 = 2^3$ and $9 = 3^2$ are consecutive square-full numbers. Now if n and $n+1$ are square-full, then so are $4n(n+1)$ and $4n(n+1)+1 = 4n^2+4n+1 = (2n+1)^2$.* □

Notice that 8 and 9 are, in fact, both perfect powers. In 1962, A. Makowski proved that there do not exist three consecutive perfect powers. However, it is an open question whether there are *any* "triples" of square-full numbers, that is, three consecutive integers all of which are square-full. How about any pairs of consecutive cube-fulls? No one yet knows, though Erdös conjectured that the answer is "no." In general, a number having prime factorization with all exponents of degree k or higher is called a *k-powerful* number, a term coined by S.W. Golomb in 1970. Our square-full numbers are thus often called *powerful* numbers in the literature. The distribution of k-powerful numbers was first studied by Erdös and Szekeres in 1935, and by many researchers ever since. In 1985, R. Heath-Brown proved that all but a finite number of positive integers can be expressed as the sum of three powerful numbers. The exceptional set is not known, however. At present the only known exceptions are 7, 15, 23, 87, 111, and 119. It has been conjectured (R. Mollin, P.G. Walsh, 1986) that these are all the exceptions. What do you think?

WORTH CONSIDERING

1. Find another point besides $p = (\sqrt{2}, \frac{1}{3})$ that works in Problem #1.
2. Modify Problem #2 to n subsets for any given $n \geq 2$.
3. **(a)** Show that there are an infinite number of positive integers that are not expressible as the sum of three nonzero squares.

   (b) Show that there are an infinite number of positive integers that are not expressible as the sum of four nonzero squares.
4. Show that for $k \geq 5$, all but a finite number of positive integers are sums of precisely k nonzero squares.
5. Show that every integer greater than 33 can be written as the sum of distinct triangular numbers (H.E. Richert, 1949).
6. Given a set of 2,005 natural numbers, show that there is a subset whose sum is divisible by 2,005.

7. Show that if nine points are placed on or in a cube having sides of length 2, then there must be two points that are at most $\sqrt{3}$ units apart.
8. Verify that 12,167 and 12,168 are a pair of consecutive numbers, with one being 2-powerful and the other 3-powerful. (Other than 8 and 9, this is the only other known such example.)

13 Primality Testing Below a Quadrillion

Carl Friedrich Gauss exclaimed, "The problem of distinguishing prime numbers from composite numbers and of resolving the latter into their prime factors is known to be one of the most important and useful in arithmetic ... The dignity of the science itself seems to require that every possible means be explored for the solution of a problem so elegant and so celebrated." Though the substance of what Gauss said 200 years ago is as true today as it was then, great progress has been made in the field of primality testing.

Most of us are well aware that trial division by all primes at most the square root of a given test number is an accurate way to determine its primality. In addition, if the number is not prime, this ancient method due to Eratosthenes (276–190 B.C.E.) will provide us with its prime factorization. If we aren't as interested in actually factoring the given number, but rather simply want to know if it is prime or not, there are many techniques that are far more efficient than the sieve of Eratosthenes. In this chapter we will not study the latest algorithms that are the absolute fastest or most efficient for determining the primality of huge numbers. However, we will gain a solid understanding of methods that can be easily implemented for numbers of size, say, less than a quadrillion. We will then mention some of the more recent methods to gain familiarity with them. When I first gave a talk on a similar topic in 1995, I asked if the current population of the United States was prime or not. The number then was 264,323,869, a number that may have been accurate at some point in time, for a few seconds at least. In any event, we'll be able to answer that particular question shortly.

A central theorem in this area is Fermat's Little Theorem. Pierre de Fermat (1601–1665), by profession a legal counselor and jurist, was also known as the "prince of amateur mathematicians." Make sure not to mistake him with C.F. Gauss, "prince of mathematicians"—or for that matter either Prince Charles, son of Queen Elizabeth, or The Artist Formerly Known as Prince. Fermat helped create the field of analytic geometry and made profound discoveries in what is now differential and integral calculus. His contributions include finding tangents and normals to curves, rectification of several plane curves, finding areas of plane

Mathematical Journeys, by Peter D. Schumer
ISBN 0-471-22066-3

regions and volumes of various three-dimensional solids. But the work he cherished most was in number theory where his discoveries were just as astonishing and penetrating. His discoveries were announced to friends by letter, but proofs generally were not as forthcoming. Many of his pronouncements waited a hundred years or more to be reproved (or disproved) by later generations. Fermat's Little Theorem is a prime example of one such result.

Theorem 13.1 (Fermat's Little Theorem): *Let p be prime and suppose that $p \nmid b$. Then $b^{p-1} \equiv 1 \pmod{p}$.*

Fermat's Little Theorem was stated by Fermat in a letter to his colleague, Frenicle de Bessy, in 1640. The moniker Fermat's Little Theorem is intended to distinguish this theorem from the perhaps more famous (but less useful) Fermat's Last Theorem. Euler generalized Theorem 13.1 and provided the first published proof in 1736. It is significant to note that oftentimes a mathematical observation can be better understood once it's been placed in a larger context. But before proceeding further, let's see in what way Euler generalized Fermat's result.

If n is a positive integer, define $\phi(n)$ to be the number of positive integers less than or equal to n that are relatively prime to n. The function ϕ is known as the *Euler phi function*. For example, please verify that $\phi(1) = 1, \phi(2) = 1, \phi(3) = 2, \phi(10) = 4$, and $\phi(20) = 8$ by listing all appropriate numbers relatively prime to the argument. Additionally, $\phi(p) = p - 1$ since every positive integer less than a given prime is relatively prime to it. For fun, you might want to check that $\phi(666) = 6 \cdot 6 \cdot 6$. We are now ready to state Euler's result, commonly called the Euler-Fermat Theorem.

Theorem 13.2 (Euler-Fermat Theorem): *If $\gcd(b, n) = 1$, then $b^{\phi(n)} \equiv 1 \pmod{n}$.*

For example, gcd(3, 10) = 1 and hence $3^{\phi(10)} = 3^4 \equiv 1 \pmod{10}$. In addition, 125 and 666 are relatively prime. Thus $125^{\phi(666)} = 125^{216} \equiv 1 \pmod{666}$, though this would be more difficult to check if it were not for our theorem. In addition, notice that Fermat's Little Theorem is a special case of the Euler-Fermat Theorem by simply constraining n to be prime. Hence Theorem 13.1 is a corollary to Theorem 13.2.

First, we need a preliminary result from elementary number theory. Then we will proceed to the demonstration of the Euler-Fermat Theorem.

Proposition 13.3: *If $\gcd(b, c) = 1$, and $b|cd$, then $b|d$.*

For example, gcd(10, 19) = 1. Since 10|380 and $380 = 19 \cdot 20$, it must be the case that 10|20. Proposition 13.3 is an extension of a result known as Euclid's Lemma, which appears as Proposition 30 in Book VII of the *Elements*. Euclid's Lemma states that if p is prime and $p|cd$, then either $p|c$ or $p|d$. Proposition 13.3 extends Euclid's Lemma by relaxing the hypothesis that b be prime. Even so, it is easy to see why Proposition 13.3 is true. Since b and c are relatively prime, all the prime factors of b must also appear in d and to high enough exponents so that b divides cd. Hence b divides d itself.

Proof of Theorem 13.2: *Let $r_1, \ldots, r_{\phi(n)}$ be a reduced set of residues modulo n. So every integer relatively prime to n is congruent to exactly one of $r_1, \ldots, r_{\phi(n)}$. In addition, since $gcd(b,n) = 1$, each of $br_1, \ldots, br_{\phi(n)}$ is relatively prime to n. Furthermore, if $br_i \equiv br_j \pmod n$ for some $i \neq j$, then $n|b(r_i - r_j)$. But $gcd(b,n) = 1$ implies that $n|(r_i - r_j)$ by Proposition 13.3. But then $r_i \equiv r_j \pmod n$, contrary to our assumption that $r_1, \ldots, r_{\phi(n)}$ form a reduced set of residues, necessarily pairwise relatively prime. Hence $r_1, \ldots, r_{\phi(n)}$ and $br_1, \ldots, br_{\phi(n)}$ are simply rearrangements of each other modulo n. It follows that*

$$(br_1)\bullet \ldots \bullet(br_n) \equiv r_1 \bullet \ldots \bullet r_n \pmod n.$$

Recombining,

$$b^{\phi(n)} r_1 \bullet \ldots \bullet r_n \equiv r_1 \bullet \ldots \bullet r_n \pmod n.$$

So $n|r_1 \cdot \ldots \cdot r_n \cdot (b^{\phi(n)} - 1)$. But $r_1 \cdot \ldots \cdot r_n$ is relatively prime to n. Thus, by Proposition 13.3, n divides $b^{\phi(n)} - 1$. Therefore, $b^{\phi(n)} \equiv 1 \pmod n$. □

Fermat's Little Theorem is a great aid in computation. For example, suppose that we wish to determine $2^{1,000} \pmod{13}$. To multiply 2 by itself a thousand times and then to divide that huge number by 13 to obtain its remainder would be unwieldy indeed. However, by Fermat's Little Theorem we know that $2^{12} \equiv 1 \pmod{13}$. Furthermore, $1,000 = 12 \cdot 83 + 4$, which we obtain by simply dividing 1,000 by 12. Hence $2^{1,000} = 2^{12 \cdot 83 + 4} = (2^{12})^{83} \cdot 2^4 \equiv 1^{83} \cdot 16 = 16 \equiv 3 \pmod{13}$. We've taken advantage of the fact that one to any power is still one.

Does Fermat's Little Theorem somehow give us a primality test? For example, to check if an odd integer n is prime, does it suffice to evaluate whether or not $2^{n-1} \equiv 1 \pmod n$? For example, what can we deduce about the number $n = 264{,}323{,}869$, which we mentioned at the beginning of this chapter? Though the computation is somewhat tedious, we find that in this case $2^{n-1} \equiv 214,721,964 \not\equiv 1 \pmod n$. By Fermat's Little Theorem, it follows that n must be composite. It turns out that $264{,}323{,}869 = 2,879 \cdot 91{,}811$; but factoring is another matter altogether that we'll happily avoid here. The key point is that we already knew that n was composite without having found any factors whatsoever. Let's investigate the applicability of Fermat's Little Theorem further.

We begin by noting that $2^{3-1} \equiv 1 \pmod 3$, $2^{5-1} \equiv 1 \pmod 5$, and $2^{7-1} \equiv 1 \pmod 7$ as guaranteed by Theorem 13.1. In addition, $2^{9-1} = 256 \pmod 3 \equiv 4 \pmod 9$. Hence, 9 must be composite by Theorem 13.1 as well. (Of course this is a silly way to show that 9 is a composite number.) Continuing, $2^{11-1} = 1024 \equiv 1 \pmod{11}$ and $2^{13-1} = 4096 \equiv 1 \pmod{13}$, indicating that 11 and 13 are both prime. Furthermore, $2^{15-1} = 16384 \equiv 4 \pmod{15}$, proving that 15 is indeed composite. Testing $2^{n-1} \pmod n$ in this way works fine until $n = 341$.

Notice that $2^{10} = 1024 = 3 \cdot 341 + 1$ and so $2^{10} \equiv 1 \pmod{341}$. Thus $2^{341-1} = 2^{340} = (2^{10})^{34} \equiv 1^{34} = 1 \pmod{341}$. Yet $341 = 11 \cdot 31$ is composite. This doesn't mean that Fermat's Little Theorem is somehow wrong. But it does mean that the *converse* of Fermat's Little Theorem is not true. Namely, the fact that $b^{n-1} \equiv 1 \pmod{n}$ does not imply necessarily that n is prime. Thus, Fermat's Little Theorem alone is not a primality test. It will not determine in all cases whether a given number is prime. But it is still extremely useful. Anytime $2^{n-1} \not\equiv 1 \pmod{n}$, that is, 2^{n-1} is *incongruent* to 1 modulo n, then n must be composite.

Our previous discussion leads us to the next definition. If n is a composite number relatively prime to b and $b^{n-1} \equiv 1 \pmod{n}$, then n is called a *base b pseudoprime*, denoted by psb(b). Roughly speaking, pseudoprimes behave pretty much like primes in that they obey the converse of Fermat's Little Theorem since they pass the test $b^{n-1} \equiv 1 \pmod{n}$.

By way of example, 25 is a base 7 pseudoprime, psp(7). Note that $25 = 5^2$ is composite and relatively prime to 7. Yet $7^{24} = (7^2)^{12} = 49^{12} \equiv (-1)^{12} = 1 \pmod{25}$. In addition, we have already shown that 341 is a psp(2).

Base 2 pseudoprimes are usually referred to as simply *pseudoprimes*, being the archetype for the other classes of pseudoprimes to other bases. For this reason, they have been studied the most carefully. Here is a chart that shows the preponderance of pseudoprimes versus actual primes up to ten billion.

Let $\pi(x)$ be the number of primes less than or equal to x and let $P\pi(x)$ be the number of pseudoprimes less than or equal to x. The function $\pi(x)$ has been calculated accurately for some enormous values of x, for example, M. Deleglise and J. Rivat have determined that $\pi(10^{18}) = 24{,}739{,}954{,}287{,}740{,}860$. In addition, R.G.E. Pinch has computed the number of pseudoprimes less than 10 trillion, specifically $P\pi(10^{13}) = 264,239$. However, we won't search nearly as far. We are more interested to get some idea of the relative frequency of pseudoprimes versus real primes as x increases. The last row of Table 13.1, $r(x)$, gives the *ratio* of integers n less than x that pass the test $2^{n-1} \equiv 1 \pmod{n}$, which are actually pseudoprimes. Specifically, $r(x) = \frac{P\prod(x)}{\prod(x)+P\prod(x)}$. This gives us some sense of the likelihood that a successful candidate is the real McCoy. The closer $r(x)$ is to zero, the more likely it is that a successful candidate is really prime.

We now have a "probabilistic" primality test. If $n < 10^{10}$ is a randomly chosen odd integer that satisfies the congruence $2^{n-1} \equiv 1 \pmod{n}$, then the probability that n is prime is greater than 0.999967. This is purer than Ivory soap! Because of the very high likelihood that a successful candidate is indeed prime, integers passing this test have been called "industrial grade" primes.

TABLE 13.1 Ratio of pseudoprimes to actual primes up to 10^{10}

x	10^2	10^3	10^4	10^5	10^6	10^7	10^8	10^9	10^{10}
$\pi(x)$	25	168	1,229	9,592	78,498	664,579	5,761,455	50,847,534	455,052,512
$P\pi(x)$	0	3	22	78	245	750	2,057	5,597	14,887
$r(x)$	0	0.017543	0.017586	0.008066	0.003111	0.001273	0.000357	0.000110	0.000033

But mathematicians, being the purists that they are, soon wondered whether there was some base b other than 2 that works better—perhaps even all the time. Unfortunately, the answer is no. In 1903, a mathematician by the name of E. Malo proved that there are infinitely many pseudoprimes (to base 2) and the following year M. Cipolla extended the result to any given base. So no matter which base b we choose, we can never make a complete list of pseudoprimes to that base.

Okay then, what else can we try? Maybe the "compositeness" of a given integer n will always be revealed if we evaluate b^{n-1} (mod n) for a handful of different b's until we find one for which $b^{n-1} \not\equiv 1 \pmod{n}$. That is, n can run, be he can't hide. For example, although $2^{340} \equiv 1 \pmod{341}$ and so 341 is a psp(2), it happens that $3^{340} \equiv 56 \not\equiv 1 \pmod{341}$. Hence 341 is not a psp(3). We would thus discover that 341 is composite if we hadn't already known. Sounds promising? Guess what, even this general notion fails.

There are composite integers n that are pseudoprimes to all bases b relatively prime to n. The smallest example isn't even that large a number. The number $n = 561$ is one such number. Let's demonstrate why this is so.

The integer $561 = 3 \cdot 11 \cdot 17$ is certainly composite. Now suppose that b is any natural number relatively prime to 561, hence necessarily relatively prime to each of 3, 11, and 17. Let's compute b^{561-1}(mod p) for p equalling 3, 11, and 17 in turn. By Fermat's Little Theorem (our handy little theorem), $b^{560} = (b^2)^{280} \equiv 1^{280} = 1 \pmod{3}$, $b^{560} = (b^{10})^{56} \equiv 1^{56} = 1 \pmod{11}$, and $b^{560} = (b^{16})^{35} \equiv 1^{35} = 1 \pmod{17}$. Since 3, 11, and 17 are pairwise relatively prime, it follows that $3 \cdot 11 \cdot 17 | (b^{560} - 1)$. Equivalently, $b^{560} \equiv 1 \pmod{561}$. Therefore, 561 is a psp(b) for any one of the infinitely many b relatively prime to 561.

The American mathematician R.D. Carmichael (1879–1967) made a careful study of such numbers. In fact, we now define a *Carmichael number* to be an odd composite n which is a psp(b) for all b relatively prime to n. Carmichael himself exhibited the first 15 such examples. Not to be outdone, here are the first 16: 561, 1,105, 1,729, 2,465, 2,821, 6,601, 8,911, 10,585, 15,841, 29,341, 41,041, 46,657, 52,633, 62,745, 63,973, and 75,361. The number 1,729 stands out as the number made famous by G.H. Hardy in reference to Ramanujan. Hardy wrote, "It was Littlewood who said that every positive integer was one of Ramanujan's personal friends. I remember going to see him once he was lying ill in Putney. I had ridden taxicab number 1729, and remarked that the number seemed to me rather a dull one, and that I hoped that it was not an unfavorable omen. 'No,' he replied, 'it is a very interesting number, it is the smallest number expressible as the sum of two cubes in two different ways.'" Indeed, $1,729 = 10^3 + 9^3 = 12^3 + 1^3$, and no smaller number has such a property. But what neither Hardy nor Ramanujan noted was that 1,729 is also a Carmichael number!

Robert Daniel Carmichael was himself an accomplished and versatile mathematician. He grew up in Alabama, the eldest of 11 talented children. After receiving his Ph.D. at Princeton under George Birkhoff in 1911, he held several posts, eventually spending the bulk of his career as longtime chair and then dean at the University of Illinois. His scientific and mathematical writings include the areas of relativity theory, differential equations, group theory, number theory, and Diophantine equations.

A very pretty result stated by O. Korselt in 1899 and rediscovered and proved by R.D. Carmichael in 1912 can be formulated as follows:

Proposition 13.4: *The number n is a Carmichael number if and only if n is a product of three or more distinct odd primes where $(p-1)|(n-1)$ for all primes p dividing n.*

By way of illustration, $1{,}729 = 7 \cdot 13 \cdot 19$. Note that $6|1{,}728$, $12|1{,}728$, and $18|1{,}728$. Thus, 1,729 satisfies the conditions of Proposition 13.4 and hence 1,729 is a Carmichael number. But how many Carmichael numbers are there? If there were just finitely many, then perhaps a list of them could be made and a primality test could still be pieced together by referring to this list together with an appropriate number of pseudoprimality tests to various bases. Let CN(x) represent the number of Carmichael numbers less than or equal to x. Table 13.2 presents a brief chart.

In 1990, Gerhard Jaeschke determined that $\mathrm{CN}(10^{12}) = 8{,}238$ as opposed to $\pi(10^{12}) = 37{,}607{,}912{,}018$. More recently, R.G.E. Pinch has computed $\mathrm{CN}(10^{15}) = 105{,}212$ and $\mathrm{CN}(10^{16}) = 246{,}683$. This compares with $\pi(10^{15}) = 29{,}844{,}570{,}422{,}669$ and $\pi(10^{16}) = 279{,}238{,}341{,}033{,}925$. So the prevalence of Carmichael numbers among the integers seems very rare as compared with the primes. But how prevalent are they?

The answer to that question was finally disposed of by R. Alford, A. Granville, and C. Pomerance in 1992. There are *infinitely* many Carmichael numbers! In fact, there are plenty of them. Alford, Granville, and Pomerance proved that there is a constant N_0 such that for any $N > N_0$, there are at least $N^{2/7}$ Carmichael numbers less than N.

Despite all this seemingly bad news, there is also plenty of good news for those of us looking for a completely reliable primality test. Here is a nice result due to the French mathematician Edouard Lucas (1842–1891), published in 1891. You may be familiar with his perennially popular children's toy, the Tower of Hanoi puzzle.

Proposition 13.5 (Lucas's Primality Test): *Given an odd integer n, if there is a base b for which $b^{n-1} \equiv 1 \pmod{n}$, but for all primes p dividing $n-1$, $b^{(n-1)/p} \not\equiv 1 \pmod{n}$, then n is prime.*

Lucas's Primality Test is sort of a "partial converse" to Fermat's Little Theorem. It has many attributes. It is a bona fide primality test rather than a probabilistic

TABLE 13.2 Abundance of Carmichael numbers up to 10^{10}

x	10^2	10^3	10^4	10^5	10^6	10^7	10^8	10^9	10^{10}
CN(x)	0	1	7	16	43	105	255	646	1,547

one. It is an "all-purpose" primality test in that it can be applied to any integer, not just to a limited set of some special form. And its implementation is fast. In fact, when it works it's a polynomial-time algorithm in the language of computational number theory and computer science. Namely, the number of steps needed to apply Lucas's Primality Test grows no faster than some polynomial with argument the input size (number of bits) of n.

But there are some obvious drawbacks as well. Here are two significant problems: One, we must first find an appropriate b, a task that is not guaranteed to be completed no matter how long we look. And two, we must be able to completely factor $n-1$. Oftentimes that's pretty easy, especially if n is known to be of some special form. However, other times knowing the factorization of $n-1$ may be just as remote.

Even so, here's an example. We will prove that $n = 7,919$ is a prime. After some work, the number $7,919 - 1 = 7,918$ can be factored as $2 \cdot 37 \cdot 107$. Lucas's Primality Test does not work on this n for $b = 2, 3$, or 5. However, next we compute $7^{(n-1)/p} \pmod{n}$ for $p = 2, 37$, and 107 in turn. The results are that $7^{(n-1)/2} \equiv -1 \pmod{n}$, $7^{(n-1)/37} \equiv 755 \pmod{n}$, and $7^{(n-1)/107} \equiv 5,549 \pmod{n}$. Since $7^{(n-1)/p} \pmod{n}$ was never one and $7^{n-1} \equiv 1 \pmod{n}$, the number $n = 7,919$ must be prime.

Drawing on ideas from Lucas's Primality Test, we can proceed further. Let p be a prime and b an integer for which $p \nmid b$. Fermat's Little Theorem implies that $p|(b^{p-1}-1)$. But $b^{p-1}-1 = (b^{(p-1)/2}+1)(b^{(p-1)/2}-1)$. By Euclid's Lemma, either $b^{(p-1)/2} \equiv 1 \pmod{p}$ or $b^{(p-1)/2} \equiv -1 \pmod{p}$. If $b^{(p-1)/2} \equiv 1 \pmod{p}$ and $4|(p-1)$, then again either $b^{(p-1)/4} \equiv 1 \pmod{p}$ or $b^{(p-1)/4} \equiv -1 \pmod{p}$. If $b^{(p-1)/4} \equiv 1 \pmod{p}$ and $8|(p-1)$, then either $b^{(p-1)/8} \equiv 1 \pmod{p}$ or $b^{(p-1)/8} \equiv -1 \pmod{p}$, and so on. This process terminates when either the modulus is $-1 \pmod{p}$ or we have exhausted all factors of 2 in the number $p-1$.

For example, consider the prime $p = 1,951$. In this case, $p - 1 = 1,950$ and $2^{1,950} \equiv 1 \pmod{1,951}$. Next $(p-1)/2 = 975$ and $2^{975} \equiv 1 \pmod{1,951}$. The process stops here since 1,975 is odd. Consider a second example, the Carmichael number $n = 561$. Remember that 561 is not prime, but behaves very much like one. In this case, $n - 1 = 560$ and necessarily $2^{560} \equiv 1 \pmod{561}$ because 561 is a Carmichael number. Next $(n-1)/2 = 280$ and $2^{280} \equiv 1 \pmod{561}$ as well. Continuing this process, $(n-1)/4 = 140$. This time $2^{140} \equiv 67 \pmod{561}$. Since $67 \not\equiv \pm 1 \pmod{561}$, it must be case that 561 is composite. Amazingly, even this Carmichael number, which can cleverly double as a prime for every one of the pseudoprimality tests, was unable to fool this seemingly simple test.

The details of this primality test were formalized by Gary Miller in 1976.

Miller's Test: Given a number n whose primality we wish to check, let $n-1 = 2^r m$ where $r \geq 1$ and m is odd. Let b and n be relatively prime. If either $b^m \equiv \pm 1 \pmod{n}$ or $b^{2^k m} \equiv -1 \pmod{n}$ for some k with $1 \leq k \leq r$, then we say that n *passes Miller's test to base b*.

Notice that, unlike Lucas's Primality Test, in Miller's Test we don't need to completely factor $n-1$. This is a big virtue. Also note that if p is prime, then p

will pass Miller's Test for any b relatively prime to p. It follows that if n fails Miller's Test for some b, then n must be composite.

If n is an odd composite that passes Miller's Test to base b, then we say that n is a *base b strong pseudoprime*, denoted by spsp(b). Strong pseudoprimes base 2 are referred to simply as *strong pseudoprimes*. It should be mentioned that if n is a spsp(b), the n is a psp(b). So being a strong pseudoprime really is more stringent than "just" being a pseudoprime.

As it happens, there are no strong pseudoprimes below 2,046. Consequently, Miller's Test correctly identifies the primality of all calendar years (so far). That is, if n is number of any year so far (C.E. = A.D.) and n passes Miller's Test, then n is in fact prime. But, unfortunately perhaps, the set of strong pseudoprimes is not empty. The first example is the number 2,047.

The number 2,047 is an interesting number, being a Mersenne number (see Chapter 5). In fact, $2{,}047 = 2^{11} - 1$. However, 2,047 is not a Mersenne prime since it's composite. In particular, $2,047 = 23 \cdot 89$. But for now, let's pretend that we didn't know the nature of the number 2,047. To apply Miller's Test, we rewrite $2{,}047 - 1 = 2{,}046$ as $2 \cdot 1{,}023$. We must check the value of $2^{1{,}023}$ (mod 2,047). In this case, there's a bit of a short cut. Since $2^{11} = 2{,}048 \equiv 1 \pmod{2,047}$ and since $1,023 = 11 \cdot 93$, it follows that $2^{1{,}023} = (2^{11})^{93} \equiv 1 \pmod{2,047}$. Hence 2,047 passes Miller's Test to base 2. But since 2,047 is composite, it is a strong pseudoprime, in fact the smallest.

It's natural to wonder, how many strong pseudoprimes are there? Table 13.3 provides a partial answer. Naturally, we let $\mathrm{SP}\pi(x)$ be the number of strong pseudoprimes less than or equal to x. If we compare Table 13.3 with Table 13.1, it is clear that there are far fewer strong pseudoprimes than there are primes (and necessarily fewer strong pseudoprimes than ordinary pseudoprimes). So Miller's Test to base 2 is an excellent tool as a probabilistic primality test. For example, the probability that a randomly chosen odd integer less than 10^{10} that passes Miller's Test to base 2 is indeed prime equals $\frac{455{,}052{,}511}{455{,}052{,}511+3{,}291} > 0.99999$. That is, if such a number n passes just this single Miller's Test, then the chance that n is prime is greater than 99.999%.

If we compare Table 13.3 with Table 13.2, to the level that we have checked, it appears that there are somewhat more strong pseudoprimes than there are Carmichael numbers. Still, the paucity of strong pseudoprimes is somewhat remarkable given that we just work with base 2 rather than with all bases as in the case of Carmichael numbers. Since our goal is to isolate the primes precisely, we might hope that there is no overlap between Carmichael numbers and strong pseudoprimes. If this were the case, then we could combine Proposition 13.4 with Miller's Test base 2 and be done. As you've probably guessed, in fact

TABLE 13.3 Abundance of strong pseudoprimes up to 10^{10}

x	10^2	10^3	10^4	10^5	10^6	10^7	10^8	10^9	10^{10}
$\mathrm{SP}\pi(x)$	0	0	5	16	46	162	488	1,282	3,291

there are numbers that are both Carmichael numbers and strong pseudoprimes. The smallest example is the number 15,841. Even so, could it still be that there are only finitely many strong pseudoprimes or finitely many strong pseudoprimes to base b for some b? The following result is a clear "no."

Proposition 13.6 (C. Pomerance, J. Selfridge, S. Wagstaff, 1980): *There are infinitely many strong pseudoprimes base b for any base b.*

Despite the fact that there are infinitely many strong pseudoprimes to any base, it has been shown that there are no odd composite integers n that are strong pseudoprimes to all bases relatively prime to n. Thus, there is no such thing as a "strong Carmichael number." In addition, M.O. Rabin has proven that if n is an odd composite, then n will pass Miller's Test for at most $(n-1)/4$ bases b with $1 < b < n-1$. So the fact that n is composite will eventually be discovered. In practice, a random selection of bases is chosen to test the primality of a candidate. Such a Miller-Rabin Test works well in practice, with demonstrably high probability. In fact, from the combined work of Miller (1976) and Bach (1985), subject to an unproved but widely believed hypothesis concerning the zeros of zeta functions known as the Generalized Riemann Hypothesis (GRH), we can be much more precise. Assuming GRH, if n is a spsp(b) for all $b \leq 2(\log n)^2$, then n is prime.

In 1980, Pomerance, Selfridge, and Wagstaff verified the following:

1. The smallest integer that is a spsp(b) for $b = 2$ is 2,047.
2. The smallest integer that is a spsp(b) for $b = 2$ and 3 is 1,373,653.
3. The smallest integer that is a spsp(b) for $b = 2$, 3, and 5 is 25,326,001.
4. The smallest integer that is a spsp(b) for $b = 2$, 3, 5, and 7 is 3,215,031,751.
5. The only odd composite below $2.5 \cdot 10^{10}$ that is a spsp(b) for $b = 2$, 3, 5, and 7 is the number 3,215,031,751.

These calculations were extended by Gerhard Jaeschke in 1993. He found that

6. The smallest integer that is a spsp(b) for $b = 2$, 3, 5, 7, and 11 is 2,152,302,898,747.
7. The smallest integer that is a spsp(b) for $b = 2$, 3, 5, 7, 11, and 13 is 3,44,749,660,383.
8. The smallest integer that is a spsp(b) for $b = 2$, 3, 5, 7, 11, 13, and 17 is 341,550,071,728,321. It is also a spsp(19), but not a spsp(23).

By combining the works cited so far, we have the following useful result for checking the primality of any number less than 300 trillion:

Proposition 13.7 (Miller-Jaeschke Primality Test): *If n is an odd integer less than $3 \cdot 10^{14}$ and n passes Miller's Test for bases $b = 2, 3, 5, 7, 11, 13$, and 17, then n is prime.*

Let n be an odd composite number. An integer b relatively prime to n is called a *witness* (to the fact that n is composite) if n is not a spsp(b). For example, the integer $n = 25{,}326{,}001$ may masquerade as a prime as far as Miller's Test is concerned for bases 2 and 3, but 5 is a witness to the fact that that n is composite. Once you've been caught, there's no denying it.

Perhaps with enough computational power, a long enough list of bases could be made so that every composite could be identified by some base from among the finite list. The bases 2, 3, 5, 7, 11, 13, and 17 catch all composites less than 300 trillion. Could it be that the set of bases consisting of say the first thousand primes would be sufficient to check any number? Again, you better take a seat before I give you the disappointing news. In 1994, W.R. Alford, A. Granville, and C. Pomerance proved that there are odd composites having least witness arbitrarily large. Hence for any finite list B of bases there are odd composites that pass Miller's Test for all b in B.

Primality testing is a very active area of current mathematical research. What we have discussed in this chapter admittedly just scratches the surface. For the sake of giving some sense of the breadth of this field, I'll simply name some important primality tests without giving any details. There are many special purpose primality tests that work specifically on numbers of a special form, for example, Mersenne numbers $2^p - 1$ (Lucas-Lehmer Test) or Fermat numbers $2^{2^n} + 1$ (Proth's Test). There have also been tremendous progress on general purpose primality tests applicable to all numbers and quite effective for testing the primality of numbers up to about 100 digits. These include the quadratic sieve and number field sieves of Carl Pomerance, the quadratic Frobenius primality test, and even the use of elliptic and hyperelliptic functions. On a theoretical level, however, these algorithms grow exponentially and so could become infeasible as the size of the input gets really large.

A major theoretical, if not actually practical, breakthrough was made in 2002 by a team of computer scientists at the Indian Institute of Technology in Kanpur. The researchers M. Agrawal, N. Kayal, and N. Saxena developed an algorithm that tests for primality in *polynomial* time! A number n has approximately log n digits. Their algorithm has been proven to have running time of order $\log^{12} n$, and may run in order of $\log^3 n$ subject to unproven but widely believed number theoretic hypotheses. Even though the algorithm may not yet be of significant aid for the numbers typically used in practice, it is a very significant, clever, and elegant addition to the field of primality testing.

Since we began this chapter with Fermat's Little Theorem, it's fitting to end with some further comments on it. If p is a prime number and b is any number with $1 \leq b \leq p-1$, Fermat's Little Theorem says that $b^{p-1} \equiv -1 \pmod{p}$. So by adding all of them, we obtain

$$1^{p-1} + 2^{p-1} + \ldots + (p-1)^{p-1} \equiv p - 1 \equiv -1 \pmod{p}.$$

In 1950, the Italian mathematician, Giuseppe Giuga, asked whether the converse is true. Namely, given an odd number n, if it turns out that $1^{n-1} + 2^{n-1} + \ldots + (n-1)^{n-1} \equiv -1 \pmod{n}$, must n be a prime? Guiga conjectured that the converse is true and he verified the truth of the converse for n up to $10^{1,000}$. If so, then

Guiga's conjecture would be another bona fide primality test, though admittedly more of aesthetic rather than practical value. In 1985, E. Bedocchi extended the verification to $10^{1,700}$. More recently, in 1996 D. Borwein, J. Borwein, P. Borwein, and R. Girgensohn have extended this bound to $10^{13,887}$. These bounds are the result of a wonderful cooperation between mathematicians and machines. Modern computers can calculate at awe-inspiring speed, but the basis for those calculations and the development of efficient uses of the computer remain the domain of human beings.

Recall that n is a Carmichael number if and only if for all p dividing n, that $p-1$ divides $n-1$. Similarly, we define n to be a *Giuga number* if for all p dividing n, that p also divides the number $\frac{n}{p}-1$. The smallest example is the number 30. Guiga proved that if n is a counterexample to his conjecture, then n must be both a Carmichael number and a Guiga number. Such a number would necessarily have at least eight prime factors. Guiga's conjecture remains open. Progress could be made by anyone. Care to try?

WORTH CONSIDERING

1. Evaluate the Euler phi function, $\phi(n)$, for the following values of n: 3, 5, 15, 24, 48, 101, 105, 1,000.

2. **(a)** Show that the function ϕ is multiplicative, that is, if gcd $(m, n) = 1$, then $\phi(mn) = \phi(m)\phi(n)$.

(b) Show that $\phi(p^t) = p^{t-1}\ (p-1)$ for p prime.

(c) Conclude that if $n = \prod_{i=1}^{m} p_i^{t_i}$, then $\phi(n) = \prod_{i=1}^{m} p_i^{t_i-1}\ (p_i - 1)$.

(d) How many positive integers one billion or less are not relatively prime to one billion?

3. Use the Euler-Fermat Theorem to show that is n is relatively prime to 10, then there is an integer consisting solely of a string of 9's that is a multiple of n.

4. Show that $\sum_{d|n}^{n} \phi(d) = n$.

5. **(a)** Factor the Carmichael number 15,841.

(b) Verify that 41,041 is a Carmichael number. In fact, it is the smallest such number consisting of four distinct prime factors.

6. **(a)** Apply Lucas's Primality Test to $n = 67$.

(b) What does Lucas's Primality Test say about the number $n = 91$?

7. Verify that 1,105 is a psp(2) and psp(3). (It is the smallest such example.)

8. C. Malo's proof of infinitely many pseudoprimes (1903):

(a) Let r be a pseudoprime and let $r' = 2^r - 1$. Show that r' is composite.

(b) Show that $2^{r'-1} \equiv 1 \pmod{r'}$

(c) Conclude that there are infinitely many pseudoprimes.

14 Erdös Number Zero

Paul Erdös

There once lived a man who ate, drank, and breathed mathematics all day long, day in and day out, year after year. He thought only about mathematics—solving interesting problems, challenging others with countless new arcane puzzles, listening and sharing his thoughts with anyone who was interested (and even with those who were not). He had almost no personal possessions and no permanent residence. For nearly 60 years, he traveled from city to city and continent to continent sharing his love of mathematics, inspiring countless young mathematicians, and collaborating with hundreds of fellow practitioners. In all he wrote

Mathematical Journeys, by Peter D. Schumer
ISBN 0-471-22066-3

or co-wrote over 1,500 mathematical articles. He walked among us until quite recently. Here is a bit of his story.

Paul Erdös was born in Budapest, Hungary, on March 26, 1913, the son of math and physics teachers, Anna and Lajos Erdös. Paul's two older sisters died of scarlet fever while his mother convalesced in the hospital following his birth. This family tragedy resulted in a very close but overly protective home environment where Paul was home schooled while his mathematical genius quickly flowered. Unfortunately, it also resulted in a socially awkward and eccentric individual who depended heavily on the care and goodwill of his friends and colleagues.

Like the mathematical giant Carl Friedrich Gauss, Erdös's mathematical talents blossomed early. At age three Paul discovered negative numbers when he correctly subtracted 250 from 100. By age four he could multiply two three-digit numbers in his head. He often entertained family friends by asking them for their birthday and then telling them the day of the week on which it occurred. To top that, he would then let them know how many seconds they had been alive!

At a young age Paul's father taught him two theorems about primes: (i) that there are infinitely many primes (Theorem 1.1), while at the same time (ii) there are arbitrarily large gaps between successive primes (Theorem 1.2). To Paul, the results seemed almost paradoxical, but they led to a deep fascination with the prime numbers and to a quest for a better understanding of their complicated arrangement. Again, like Gauss, an early fascination with number theory was the impetus for Erdös's lifetime dedication (some might say addiction) to the world of mathematics.

Another early mathematical influence was the *Hungarian Journal for Secondary Schools*, commonly called KoMal. The most popular part of the journal was a regular problem section where student solutions were published and individual credit was given for correct solutions. At the end of the year the pictures of the most prolific problem solvers were included in the journal. In this way, the most ingenious mathematics and science students were introduced to one another. In this sense, Paul's "publications" began when he was barely 13 years old.

Erdös's first significant result was a novel elementary proof of Bertrand's Postulate. In 1845, the French mathematician Joseph Bertrand (1822–1900), a child prodigy himself who was attending lectures at the Ecole Polytechnique at age 11, conjectured that for any natural number $n > 1$, there was always a prime between that number n and its double $2n$. Bertrand himself verified the conjecture for all n up to three million, but he was unable to prove it in general. Five years later, the Russian Pafnuti Lvovich Chebyshev (1821–1894) proved the result, though the name Bertrand's Postulate seems to have stuck in perpetuity. Chebyshev's proof was brilliant, but it was also difficult and relied heavily on analytic methods (advanced calculus). At age 18, Erdös created a wholly new proof, which, though quite intricate, was elementary in the sense that no calculus or other seemingly superfluous analytical methods were employed. In fact, this proof together with other related results on primes in various arithmetic progressions constituted his doctoral dissertation.

Another early success was a generalization of a fascinating observation by one of his close friends. Esther Klein noticed and proved that for any five points in the plane, no three collinear, it must always be the case that four of them can be chosen, forming the vertices of a convex quadrilateral. (A convex shape is one in which if any two points within it are chosen, then the line joining those two points lies entirely within the shape itself.) Paul Erdös and George Szekeres were able to extend this to a more general result. They demonstrated that for any number n there is a corresponding number N, so that any N points in the plane (with no three collinear) have a subset of n points that form a convex n-sided polygon. Since Esther and George became romantically involved during this period and later married, the result was dubbed the Happy End Problem. Furthermore, Erdös and Szekeres conjectured that in fact the smallest such N will always be equal to $2^{n-2} + 1$. Interestingly, the more general conjecture still has not been proven. However, the Happy End Problem was a harbinger of much of Erdös's later work—fruitful collaborations, beautiful theorems, and tantalizing conjectures. A good theorem often spawns more questions than it answers.

Erdös also proved a neat result about abundant numbers. Let $s(n)$ represent the sum of all the proper divisors of n (i.e., all positive divisors except n itself). Then n is called *deficient*, *perfect*, or *abundant*, depending on whether $s(n)$ is less than, equal to, or greater than n, respectively. Such numbers have been studied since the time of Pythagoras. (See Chapter 5 for more background.) The German mathematician Issai Schur (1875–1941) conjectured that the set of abundant numbers had positive density. That is, let $A(x)$ be the number of abundant numbers less than or equal to x, then Schur's conjecture states that $\lim_{x \to \infty} \frac{A(x)}{x}$ exists and is strictly greater than zero. Erdös gave a wonderful proof of this. When Schur heard about the success of the young Hungarian, he dubbed him "the magician of Budapest."

With life in Hungary ever worsening for Jewish intellectuals, Erdös obtained a fellowship at the University of Manchester in England. Luckily, Erdös cancelled his original plan, which was to tour Germany. So in 1934, he headed off for England, first visiting Cambridge University and then on to Manchester. Thus began his mathematical travels and worldwide collaborations, an adventure that did not abate until the day he died.

It would be 14 years before Erdös was able to return to Budapest and be reunited with his mother, who had miraculously survived the war. Unfortunately, Paul's father died during this period and four out of five of his aunts and uncles perished in the Holocaust. Paul's beloved mother spent much of the rest of her life traveling with her son and utilizing her apartment as a repository for his ever-increasing mountain of reprints.

Though Erdös's relationship with the American government was generally harmonious, things soured during the McCarthy era. In 1954 while on a temporary faculty position at Notre Dame, Erdös wished to attend the International Congress of Mathematicians being held in Amsterdam. Knowing that he came from a Communist country, an agent from the Immigration and Naturalization Service interviewed Erdös and asked him what he thought of Karl Marx. Erdös replied forthrightly, "I'm not competent to judge. But no doubt, he was a great man." Perhaps due to this incident, Erdös was denied a re-entry visa to the United States after attending the Congress. Fortunately, strong support and numerous letters to senators and congressmen finally resulted in Erdös being allowed to return to the United States in 1959. From that point on, he could come and go freely—and he certainly did.

To all who knew Erdös, it appeared as though he spent 99 percent of his wakeful hours obsessed with doing mathematics (although he somehow developed a deceptive skill at both table tennis and the game of go). Twenty hours of work a day was not at all unusual. Upon arriving at a professional meeting, he would announce in his thick accent, "My brain is open." At parties, he would often stand apart, deep in thought pondering some abstruse mathematical argument. When being introduced to a math graduate student still struggling to complete a dissertation, Erdös would typically ask, "What's your problem?" One would normally be taken aback by such a remark if it were uttered by a stranger in less friendly surroundings, but with Erdös it was clearly meant as a genuine open-armed welcome. It meant he took you seriously as a fellow dweller in his mathematical world.

One of Erdös's greatest triumphs was his elementary proof of the Prime Number Theorem (PNT). The PNT describes the asymptotic distribution of the prime

George Bernhard Riemann

numbers and variants of it were conjectured by both Gauss and Legendre in the late 1700s. Specifically, let $\pi(x)$ be the number of primes less than or equal to x and let $\text{Li}(x) = \int_2^x \frac{1}{\log t} dt$ where log is the natural logarithm function. The PNT states that the $\lim_{x \to \infty} \frac{\pi(x)}{Li(x)} = 1$, that is, the number of prime less than x is asymptotic to $\text{Li}(x)$.

Significant progress toward a proof of the PNT was made by Chebyshev in the 1850s and by George Bernhard Riemann (1826–1866) in 1859. Riemann's contribution was based on a deep and careful study of the complex-valued zeta function, now celebrated as the Riemann zeta function. Finally, in 1896 the French mathematician Jacques Hadamard (1865–1963) and the Belgian mathematician Charles Jean de la Vallée Poussin (1866–1962) each independently proved the result using delicate arguments from complex function theory. Indeed many feel that the proof of the PNT was the mathematical capstone of the nineteenth century.

In the first half of the 20th century, the search for an "elementary" proof of the PNT seemed completely hopeless. However, in 1949 Paul Erdös and the Norwegian mathematician Atle Selberg, working in tandem but not together, found such proofs. Selberg appropriately won a Fields Medal for his work while Erdös won the prestigious Cole Prize in algebra and number theory for his contribution.

Another area that fascinated Erdös was classical Ramsey theory, which describes the number of ways of partitioning a set into a given number of subsets under certain particular constraints. The party problem is the classic example: at least how many people must there be so that there are either three mutual friends or three mutual strangers at a party? Here we make the amicable assumption that whoever you know is a friend. (Equivalently, we could discuss mutual acquaintances and nonacquaintances.)

A little graph theory background is helpful. A *graph* G consists of a set of vertices together with some set of lines, each of which adjoin two vertices. Vertices are said to be adjacent if there is a line adjoining them. Furthermore, the *degree* of a vertex is the number of lines emanating out from that vertex, equivalently it's the number of adjacent vertices. Given a graph G, a closely related graph is its *complementary graph*, $\overline{G}$. The complementary graph has the same vertex set as G. However, two vertices are adjacent in $\overline{G}$ if and only if they are not adjacent in G. Finally, define the *complete graph* on n vertices to be the graph K_n, which consists of n vertices each of degree $n-1$, that is, every vertex is connected to every other one.

We can represent the relations among people at a party by a graph. Each vertex stands for an individual party-goer. If two people are friends, then they are joined by a line. If they are strangers, then no line is drawn. In Figure 14.1*a* we have drawn a "friendship" graph G for a party involving five people. Notice that there are no triangles, and hence no set of three mutual friends. (Not much of a party, huh?) The graph $\overline{G}$ is shown in Figure 14.1*b*. Again there is no triangle. Hence there is no set of three mutual strangers at the party either. However, with six people present at the party the situation changes.

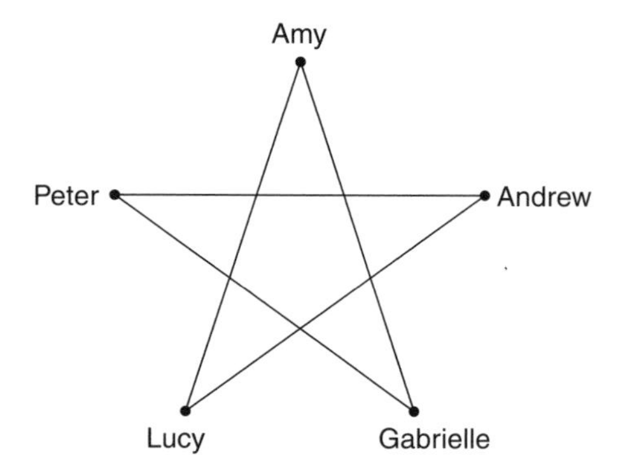

Figure 14.1a Five person friendship graph G with no triangle.

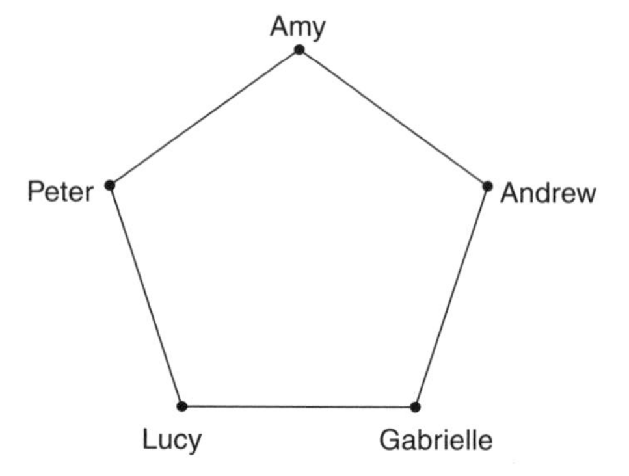

Figure 14.1b Complementary graph $\overline{G}$.

Theorem 14.1: *A party with six or more people will consist of a set of three mutual friends or three mutual strangers.*

Proof of Theorem 14.1: *Let G be the friendship graph for any particular party with six people and let $\overline{G}$ be its complementary graph. For any given vertex v, the sum of its degrees in G and $\overline{G}$ is five because everyone else is either a friend or a stranger. Without loss of generality, assume that the degree of v in G is at least three. Let v be adjoined to v_1, v_2, and v_3. If any of those vertices are adjacent, then there would be a triangle in G formed from v and the two other points corresponding to three mutual friends. Otherwise, v_1, v_2, and v_3 are mutually nonadjacent corresponding to three mutual strangers. To complete the proof, note that any larger party contains a subset of six people which we have shown satisfies the theorem.* □

More generally, define $r(m, n)$ to be the smallest integer r such that if the edges of a complete graph on r vertices are colored one of two colors, then there must be a complete subgraph on m vertices of one color or a complete subgraph of

n vertices of the other color. So Theorem 14.1 can be rephrased to assert that $r(3, 3) = 6$. The values of $r(3, n)$ are also known for $n = 4, 5, 6, 7$, and 9. In addition, with significantly more work it has been shown that $r(4, 4) = 18$. So a party with 18 or more people must contain a group of four mutual friends or four mutual strangers. Erdös proved that $r(m, n) \leq \frac{(m+n-2)!}{(m-n)!(2n-2)!}$, the number of combinations of $m + n - 2$ objects chosen $m - n$ at a time. Furthermore, Erdös offered \$250 to anyone who could prove that $\lim_{n\to\infty} r(n, n)^{1/n}$ exists. If the limit does exist, it is known to be between $\sqrt{2}$ and 4. Interestingly, the value of $r(5, 5)$ is still unknown, though it must be at least 43 and at most 49. Erdös was fond of saying that if an evil spirit was going to destroy the world unless $r(5, 5)$ could be determined, then it would be prudent for all nations to devote all its resources to this problem. On the other hand, if the evil spirit insisted on knowing $r(6, 6)$, then it would make better sense to devote all our resources to destroy the evil spirit!

Erdös not only loved working on difficult problems and proving challenging theorems, but he always strived for the most elegant and direct proof. He had unusual religious views and referred to the Almighty as the SF (for Supreme Fascist). Erdös felt that he was forever in the midst of an ongoing personal battle with the SF. However, one positive aspect of this was the SF kept a secret book, *The Book*, which had all the theorems that would ever or could ever be discovered along with the simplest and most elegant proofs for each one. The highest compliment Erdös would give was that someone's proof was "one from The Book."

Here are a sampling of some his theorems, and although the proofs would be inappropriate here, many certainly would qualify as being one from The Book.

1. There are infinitely many odd integers that are not expressible as the sum of a prime number and a power of two.
2. The product of two or more consecutive positive integers is never a square or any other higher power.
3. A connected graph with minimum degree d and at least $2d + 1$ vertices has a path of length at least $2d + 1$.
4. Let $d(n)$ be the number of divisors of n. Then the series $\sum_{n=1}^{\infty} \frac{d(n)}{2^n}$ converges to an irrational number.
5. Let $g(n)$ be the minimal number of points in general position in the plane needed to ensure a subset exists that forms a convex n-gon (from the Happy End Problem). Then $2^{n-2}+1 \leq g(n) \leq \frac{(2n-4)!}{(n-2)!^2}+1$. (Fan Chung and Ronald Graham have recently removed the "+1" from the upper bound!)

To keep this chapter from being overwhelmingly expository, we can include one beautiful theorem with a truly elementary proof. The result itself deals with basic properties of finite real-numbered sequences. We begin with a question for you.

Can you somehow jumble up the sequence 1, 2, 3, 4 so that there is no 3-term increasing subsequence nor any 3-term decreasing subsequence? Note that a subsequence does not have to consist of consecutive elements of the original sequence; rather we only insist that their order remain unchanged. For example, 1, 3, 2, 4 has no 3-term decreasing sequence, but it does have 1, 2, 4 as a 3-term increasing sequence. No doubt you've discovered that 3, 4, 1, 2 (or its reverse 2, 1, 4, 3) has no 3-term subsequence that either strictly increases or decreases. There is another pair that I'll let you find for yourself. Also note that this problem made no use of the arithmetic properties of the natural numbers. It just seemed more natural to introduce the problem this way.

What about the same question for the sequence 1, 2, 3, 4, 5? Can you mess up the numbers somehow so that there is no 3-term increasing nor 3-term decreasing subsequence? After a short while, you should be amply convinced that in this case the answer is no. In fact, all rearrangements of 1, 2, 3, 4, 5 will have some 3-term subsequence that either increases or decreases. A quick way to verify this is to take all the 4-term sequences of 1, 2, 3, 4 that had no such 3-term subsequence and insert the number 5 in all possible positions (intermediate or at either end). For example, take the sequence 3, 4, 1, 2. Both sequences 5, 3, 4, 1, 2 and 3, 5, 4, 1, 2 have decreasing subsequence 5, 4, 2. In addition, the sequences 3, 4, 5, 1, 2 and 3, 4, 1, 5, 2 and 3, 4, 1, 2, 5 all have increasing subsequence 3, 4, 5. Now check this on the other cases.

Ready for a tougher challenge? Can you jumble up the sequence 1, 2, 3, 4, 5, 6, 7, 8, 9 so that there is no 4-term increasing subsequence or 4-term decreasing subsequence? There are $9! = 362{,}880$ ways to rearrange the first nine digits, but even so you may hit on a solution. If you're ready to take a look, here's one solution: 3, 7, 2, 1, 9, 5, 4, 8, 6.

Okay then, what about the sequence 1, 2, 3, 4, 5, 6, 7, 8, 9, 10? Now there are $10! = 3{,}628{,}800$ permutations and things are getting a bit unwieldy. Here's where a good mathematical theorem is needed. The following theorem was proved by P. Erdös and G. Szekeres in an article entitled, "A Combinatorial Problem in Geometry," *Compositio Math.*, 1935. As a consequence, all permutations of the numbers 1 to 10 must have either a 4-term increasing subsequence or 4-term decreasing subsequence.

Theorem 14.2: *In any sequence $a_1, a_2, \ldots, a_{mn+1}$ of $mn+1$ distinct real numbers, there always exists either an increasing subsequence of length $m+1$ or a decreasing subsequence of length $n+1$ (or both).*

Proof of Theorem 14.2: *Consider the sequence $a_1, a_2, \ldots, a_{mn+1}$. For each $a_i (1 \leq i \leq mn+1)$, let s_i be the length of the longest increasing subsequence starting at a_i. To check your understanding, here is the chart for the sequence 1, 5, 3, 4, 2:*

a_i	1	5	3	4	2
s_i	3	1	2	1	1

If $s_i \geq m+1$ for some i, then we are done since there would be an increasing subsequence of length at least $m+1$. Otherwise, $s_i \leq m$ for all i. In this case, *consider the function $f(a_i) = s_i$ defined for all i. The domain of f is the set $\{a_1, a_2, \ldots, a_{mn+1}\}$, and the co-domain of f is the set $\{1, 2, \ldots, m\}$. Recall that the co-domain of a function is the set from which the range of f must choose. No matter how equally distributed the values of f are, by the pigeonhole principle there is a value t from among $\{1, 2, \ldots, m\}$ such that $f(a_i) = t$ for at least $\frac{mn+1}{m} = n + \frac{1}{m}$ such a_i. But the number of such a_i must be an integer. Thus there is a value t for which $f(a_i) = t$ for at least $n + 1$ such arguments. In particular, let $a_{j_1}, a_{j_2}, \ldots, a_{j_{n+1}}$ with $j_1 < j_2 < \ldots < j_{n+1}$ be these numbers. For $i = 1, 2, \ldots, n$, it must be the case that $a_{j_i} > a_{j_{i+1}}$ or else there would be an increasing subsequence of length $t + 1$ beginning at a_{j_i}, contradicting the assumption that $f(a_{j_i}) = t$. But then it must be the case that $a_{j_1} > a_{j_2} > \ldots > a_{j_{n+1}}$, a decreasing subsequence of length $n + 1$.* □

Nothing was more exciting to Erdös than to discover a mathematically talented child and to excite him or her about doing mathematics. In 1959, Erdös arranged to have lunch with a very precocious 11-year old, Lajos Posá. Erdös challenged the youngster to show why if $n+1$ integers are chosen from the set $\{1, 2, \ldots, 2n\}$, then there must be two chosen numbers that are relatively prime. Clearly the set of even numbers less than or equal to $2n$ does not have this property, showing that just choosing n integers is insufficient. According to Erdös, within half a minute Posá solved the problem by making the striking observation that two consecutive integers must always be chosen and they are necessarily relatively prime. Erdös commented that rather than eating soup, perhaps champagne would have been more appropriate for this occasion!

Erdös's greatest influence on fellow mathematicians, young and old alike, was his continual outpouring of new conjectures coupled with various monetary rewards. Some problems had a price tag of just a few dollars while others went for several thousand dollars. To solve an Erdös problem is considered a great accomplishment, and the larger the reward the more difficult Erdös considered the problem to be. Erdös was often asked what would happen if all his problems were solved simultaneously. Could he possibly pay up? His answer was that of course he could not. Then he'd add, "But what would happen if all the depositors went to every bank and demanded all their money?" Certainly the banks could not pay; and yet this latter scenario is far more likely to occur.

Here are a few outstanding conjectures and open problems from Erdös:

1. Do there exist infinitely many primes p such that every even number less than or equal to $p-3$ can be expressed as the difference between two primes each at most p? For example, 13 is such a prime since $10 = 13 - 3$, $8 = 11 - 3$, $6 = 11 - 5$, $4 = 7 - 3$, and $2 = 5 - 3$. The smallest prime not satisfying this condition is $p = 97$.
2. For every integer n are there n distinct integers for which the sum of any pair is a square? For example, for $n = 5$, the numbers $-4{,}878$, 4,978, 6,903, 12,978, and 31,122 have this property.

3. Is there a polynomial $P(x)$ for which all sums $P(a) + P(b)$ are distinct for $0 \leq a < b$? For example, $P(x) = x^3$ doesn't work since $10^3 + 9^3 = 12^3 + 1^3$. However, $P(x) = x^5$ is considered to be a likely candidate.
4. Are there infinitely many primes p for which $p - n!$ is composite for all n such that $1 \leq n! \leq p$? For example, for $p = 101$, $p - n!$ is composite for $n = 1, 2, 3$, and 4. (Erdös conjectured that the statement is false.)
5. A natural number is *pseudoperfect* if it is abundant and also expressible as a sum of some subset of its proper divisors. For example, 66 is pseudoperfect since 66 is abundant and yet $66 = 11+22+33$ where all terms are divisors of 66. A number is *weird* if it is abundant but not pseudoperfect. For ten dollars, are there any odd weird numbers?
6. A system of congruences $a_i \pmod{m_i}$ for $1 \leq i \leq k$ where $m_1 < m_2 < \ldots < m_k$ is a *covering system* if every integer satisfies at least one of the congruences. For example, 0 (mod 2), 0 (mod 3), 1 (mod 4), 5 (mod 6), and 7 (mod 12) form a covering system. Given any positive integer c, is there a covering system with $m_1 \geq c$? (Currently $c = 24$ is the largest value that has been constructed.) Another open question of Erdös and John Selfridge asks whether there is a covering system with all moduli odd integers.
7. A \$5,000 conjecture: Let $A = \{a_i\}$ be any sequence of natural numbers for which $\sum_{i=1}^{\infty} \frac{1}{a_i}$ diverges. Is it true that A must contain arbitrarily long arithmetic progressions? If so, one corollary would be that the set of primes contains arbitrarily long arithmetic progressions. Currently the record is 22 primes in arithmetic progression (due to A. Moran and P. Pritchard, 1993). Additionally, a set of 10 consecutive primes in arithmetic progression is known (H. Dubner et al., 1998). Each prime is 93 digits long!

Erdös was almost as well-known for his eccentricities as he was for his brilliant mind. By his own admission, Erdös never attempted to butter his own toast until he was already an adult. "It turned out not to be too difficult," he admitted. Many mathematicians have had the experience of either walking or driving Erdös to his next commitment, only to learn after some time that he assumed you knew where he was supposed to be going.

Erdös had an aversion to old age and its concomitant infirmaries as well as an obsession with death itself. When breaking off work for the night he would say, "We'll continue tomorrow—if I live." At age 60, Erdös started appending acronyms to his name. The letters pgom stood for "poor great old man." Five years later he added ld for "living dead" and so on. Eventually he got to cd, denoting "counts dead." This was explained as follows: The Hungarian Academy of Sciences has a strict limit on the total number of members that it can have at one time. However, once you reach the age of 75, you can still remain a member but you are no longer counted among the total. Therefore, at that point, you are counted as if you were dead.

When asked, "How old are you?," Erdös would answer that he was two-and-a-half billion years old. After all, when he was a child he was taught that the earth was two billion years old and now they say that it's four-and-a-half billion years old! Some say that Erdös was the Bob Hope of mathematicians. Not only did he share his humorous stories and witticisms, but he traveled widely and through his lectures raised the morale of the mathematical troops. He called such lecturing "*preaching*," an appropriate term for someone so dedicated to the importance of spreading the good mathematical word.

The shorthand language that Erdös used is oft referred to as Erdösese. An *epsilon* was a child, *poison* meant alcoholic drink (which he scrupulously avoided), *noise* referred to music, *boss* was wife, and *slave* was husband. If someone was *captured* that meant he or she was married, while *liberated* stood for divorced. If a mathematician stopped publishing he *died*, while actually dying was referred to as *having left*. Nothing bothered Erdös more than political strictures that did not allow for complete freedom of expression and the ability to travel unhindered. The Soviet Union was called *Joe* (for Joseph Stalin) and the United States was known as *Sam*.

No account of Paul Erdös would be complete without mentioning the concept of Erdös numbers. Erdös himself had Erdös number zero. Anyone who co-authored a paper with him (there are 507 such people) have Erdös number one. Those who did not, but who co-authored a paper with Erdös number one are assigned Erdös number two (of which there are currently at least 5,019 such people), and so on. Like golf, the lower the number, the more prestigious the result. The largest Erdös number believed to exist is seven. Erdös himself added an interesting twist for those with whom he directly collaborated. Instead of all having Erdös number one, he claimed that someone co-authoring n papers should be assigned number $1/n$. The lowest Erdös number in this case would be held by Andras Sarkozy with number 1/57—just edging out Andras Hajnal with number 1/54.

Paul Erdös received countless honorary degrees and his work was and continues to be the focus of many international conferences. In 1984, Erdös received the highly prestigious Wolf Prize for his lifetime's contributions to the world of mathematics. Of the $50,000 awarded, he immediately donated $49,280 to an Israeli scholarship named in memory of his mother. On other occasions, he donated money to Srinivasa Ramanujan's widow, to a student who needed money to attend graduate school, to a classical music station, and to several Native American causes. Always traveling with a single shabby suitcase that doubled as a briefcase, he had little need or interest in the material world. He had no home and precious few possessions. Without hesitation, he once asked the versatile Canadian mathematician Richard Guy for $100 adding, "You're a rich man." Richard Guy gladly gave him the money. Later Guy poignantly noted, "Yes, I was. I knew Paul Erdös."

Paul Erdös was an extremely creative and versatile mathematician. His work spanned number theory, geometry, graph theory, combinatorics, Ramsey theory, set theory, and function theory. He helped create new areas of inquiry—probabilistic number theory, extremal graph theory, the probabilistic method, and much of what is now broadly referred to as discrete mathematics. The vast quantity of his research output alone qualifies him as the Euler of our age. He far surpassed Einstein's litmus test for success, which was to be highly esteemed by one's colleagues rather than to be popularly well-known. Once during a lecture by the late number theorist Daniel Shanks, a long computer-generated computation resulted in a 16-digit number. Shanks, who was not known for sprinkling praise lightly said, "I don't know if anyone really understands numbers like these—well, maybe Erdös."

A fantasy that Paul Erdös had was that he'd die in the midst of giving a lecture. After proving an interesting result, a voice from the audience would pipe up, "But what about the general case?" Erdös would reply, "I leave that to the next generation," and then immediately drop dead. In fact, Erdös left us on September 20, 1996, while attending a mathematics conference in Warsaw—not so different from his fantasy.

So ... was he one of the greatest mathematicians of the last century? Echoing Erdös himself, I'm not competent to judge. But no doubt he was a great man.

WORTH CONSIDERING

1. Prove that there are infinitely many primes of the form $4k+3$. *Hint*: Assume otherwise and consider the number $N = 4p_1 \cdots p_n - 1$ where $p_1, \ldots, p_n$ are all primes congruent to 3 modulo 4.

2. Verify Bertrand's Postulate for $1 < n < 1{,}000$.

3. **(a)** Verify that the number 945 is an abundant number. In fact, it is the smallest odd abundant number.

(b) Show that there are infinitely many odd abundant numbers.

4. Find all values of n for which $\sigma(n) = 12$.

5. **(a)** Show that any multiple of a perfect or abundant number must be abundant.

(b) Verify that 46 cannot be expressed as the sum of two abundant numbers.

(c) Show that any even number greater than 46 can be expressed as the sum of two abundant numbers. (*Hint*: Use the fact that 12, 20, 40, and all multiples of 6 are abundant numbers.)

6. Verify that for every prime p less than 97, every even number at most $p-3$ can be expressed as the difference between two primes, each at most p.

7. A related pair of unproven conjectures:

(a) In 1948, P. Erdös and E.G. Straus conjectured that the equation $\frac{4}{n} = \frac{1}{x} + \frac{1}{y} + \frac{1}{z}$ is solvable in positive integers for all $n > 1$. Verify the result for $n \leq 12$.

(b) In 1956, W. Sierpínski conjectured that the equation $\frac{5}{n} = \frac{1}{x} + \frac{1}{y} + \frac{1}{z}$ is solvable in positive integers for all $n > 1$. Verify this result for $n \leq 12$.

8. Rearrange the numbers 1, 2, 3, 4, 5, 6 so that there is no 4-term increasing subsequence nor 3-term decreasing subsequence.

9. Give an example of a *weird* number, namely one that is abundant but not pseudoperfect.

10. **(a)** In 1937, Erdös asked the following, "What is the largest number of elements that can be chosen from the set $S = \{1, 2, \ldots, 2n\}$ so that none divides any other?" What is the answer?

(b) What is the largest number of elements from the set $\{1, 2, \ldots, n\}$ that can be chosen so that no member divides two of the others?

15 Choosing Stamps to Mail a Letter, Let Me Count the Ways

Suppose you wish to mail a 20-cent postcard and have an inexhaustible supply of stamps of all denominations. How many choices do you have in selecting the stamps? For example, you might choose one 20-cent stamp, or you might choose three 5-cent stamps and five 1-cent stamps, or you might choose a 10-cent stamp and five 2-cent stamps, etc. In addition to the three possibilities just listed, you might be surprised to learn that there are 624 additional possibilities. That's right, there are 627 ways to choose stamps to mail a 20-cent postcard. And what about a 37-cent first-class letter? Here there is a whopping 21,637 possible selections! Maybe it's really easier to just stick with e-mail after all.

Mathematically, we don't really care about physical stamps, but rather about the number of ways to sum to a given total. In the 17th century, G.W. Leibniz fiddled around with this for awhile. He attempted to discover a formula that would give the number of ways of summing to any given natural number n. After some time, he gave up in frustration, but did write to Johann Bernoulli that the problem was certainly "difficult and interesting." The function was soon called the partition function and many mathematicians and other natural philosophers began to study it. In particular, let n be an integer. A *partition* of n is a representation of n as a sum of positive integers; the terms are called the *parts* of the partition. The total number of essentially distinct partitions is denoted by $p(n)$. In distinguishing among partitions, order doesn't matter. So $5+3+2$ and $2+3+5$ are considered to be the same partition of the number 10. Hence $p(n)$ is known as the *(unordered) partition function*.

In the 18th century, Leonhard Euler made a host of wonderful discoveries about the partition function. With his usual commanding insight, he discovered several interesting connections between partitions of different forms. Though he did not discover an actual formula for $p(n)$, he did develop a clever method of calculating the number of partitions of n based on knowing the answer for smaller arguments. In this chapter we will discuss and amplify some of Euler's discoveries. But first let's dispose of the much easier problem of counting all the *ordered* partitions. The analogous function is the composition function $c(n)$,

Mathematical Journeys, by Peter D. Schumer
ISBN 0-471-22066-3

Gotifried Wilhelm Leibniz

which stands for the number of ways of summing to n with the order of the parts counting. In this case, $5 + 3 + 2$ and $2 + 3 + 5$ would constitute two different *compositions* of the number 10. There is a straightforward method of representing compositions which will provide the solution to our problem.

Given a composition of n, we can represent it graphically by placing n dots in a row and divide them appropriately. For example, the composition of 10 given by $5 + 3 + 2$ would be drawn as

•••••|•••|••

while the composition $2 + 3 + 5$ would be represented this way:

••|•••|•••••

Each composition of 10 consisting of three parts corresponds to a diagram with ten dots and two lines placed among them. In addition, we can't place the two lines next to each other since each part must be a number greater than or equal to one. Conversely, any choice of two separated lines corresponds to a unique composition of 10. Since there are nine gaps between the ten dots and two lines to be placed, the total number of compositions of 10 consisting of three parts is the binomial coefficient $\binom{9}{2} = \frac{9!}{2!7!}$. Similarly, the number of compositions of 10 consisting of four parts is $\binom{9}{3}$, and the number of compositions of 10 consisting of n parts is $\binom{9}{n-1}$. But compositions of 10 could have any number of parts from just one (the composition 10) to ten (the composition consisting of the sum of ten ones). Therefore, the total number of compositions of 10 is $\binom{9}{0}+\ldots+\binom{9}{9}$. But this is the sum of all entries across the ninth row of Pascal's triangle corresponding to the total number of subsets of the number 9. We know this number is 2^9. Similarly, the number of compositions of the number n, $c(n)$, is 2^{n-1}.

We now return to our main concern, that of (unordered) partitions. Since the order of terms does not matter, we may assume that for any given partition that the terms are listed in nonincreasing order. For example, the partitions of 5 are 5, 4+1, 3+2, 3+1+1, 2+2+1, 2+1+1+1, and 1+1+1+1+1. A clever idea of N.M. Ferrers (1829–1908), one-time Senior Wrangler at Cambridge University, is to depict partitions visually as an array of dots, now called Ferrers's graphs. For example, the partition $3+2$ is represented as

• • •
• •

while the partition $2+1+1+1$ is given by

• •
•
•
•

A Ferrers's graph of a partition can be read from top to bottom or from left to right. Hence the first partition of 5, namely $3+2$, can also be interpreted as 2+2+1. This is known as the *conjugate* partition. Similarly, the second partition of 5, that is, $2+1+1+1$, has conjugate partition $4+1$. Thus, two immediate consequences of Ferrers's graphs are the following:

Proposition 15.1: *The number of partitions of n into k parts is equal to the number of partitions of n with largest part k.*

Proposition 15.2: *(a) The number of partitions of n into an even number of parts is equal to the number of partitions of n with largest part an even number.*

(b) The number of partitions of n into an odd number of parts is equal to the number of partitions of n with largest part an odd number.

For example, the partitions $3+2$ and $3+1+1$ are the two partitions of 5 with largest part 3. Their conjugate partitions, $2+2+1$ and $3+1+1$, are the two partitions of 5 into three parts. By the way, the partition $3+1+1$ is known as a self-conjugate partition.

Next we introduce the notion of *generating functions*, a key concept in many areas of mathematics. Recall that the geometric series $1+x+x^2+\ldots$ converges absolutely to the function $\frac{1}{1-x}$ for $|x|<1$. Similarly, the geometric series $1+x^k+x^{2k}+\ldots$ converges absolutely to $\frac{1}{1-x^k}$ for $|x|<1$. If we multiply the two series together corresponding to $k=1$ and $k=2$, then we obtain

$$\frac{1}{(1-x)(1-x^2)} = (1+x+x^2+x^3+x^4+\ldots)(1+x^2+x^4+\ldots).$$

Since each series converges absolutely we can multiply and collect like terms. We obtain

$$\frac{1}{(1-x)(1-x^2)} = 1+x+2x^2+2x^3+3x^4+3x^5+\ldots \qquad (15.1)$$

Let us denote the number of partitions of a number n with largest part at most k by $p(n, k)$. For example, the partitions of 5 with largest part at most 2 are the three partitions $2+2+1, 2+1+1+1$, and $1+1+1+1+1$. Hence $p(5,2)=3$. If we look at the coefficient of x^5 in Equation 15.1, we will see that it is 3. That is due to the fact that each occurrence of x^5 there corresponds to a sum of products of x's and x^2's. By the law of exponents, we add exponents when multiplying quantities with the same base. For example, the partition $2+2+1$ corresponds to the product $x^2 \cdot x^2 \cdot x^1$ while the partition $2+1+1+1$ corresponds to the product $x^2 \cdot x^1 \cdot x^1 \cdot x^1$. Thus the left-hand side of Equation 15.1 is a generating function for the partition function $p(n, 2)$. We can thus rewrite Equation 15.1 as follows:

$$\frac{1}{(1-x)(1-x^2)} = \sum_{n=0}^{\infty} p(n,2)x^n \qquad (15.2)$$

Here we define $p(0,k)=1$ for any $k \geq 1$. Analogously, we can extend our product to contain m factors, thereby yielding a generating formula for $p(n, m)$:

$$\frac{1}{(1-x)(1-x^2)\cdots(1-x^m)} = \sum_{n=0}^{\infty} p(n,m)x^n \qquad (15.3)$$

If we extend the product so that $k \to \infty$, then we obtain the generating function for the partition function.

Proposition 15.3 (Generating Function for the Partition Function): *Let* $P(x) = \prod_{k=1}^{\infty} \frac{1}{1-x^k}$. *Then* $P(x) = \sum_{n=0}^{\infty} p(n)x^n$ *for* $0<x<1$. *So* $P(x)$ *is the generating function for the partition function,* $p(n)$.

It's quite instructive to actually use $P(x)$ to calculate the first few terms of the series. Let's use the generating function to calculate the values of $p(n)$ up to and including $n = 6$. In order to do so we can ignore any factors beyond x^6:

$$P(x) = (1 + x + x^2 + x^3 + x^4 + x^5 + x^6 + \ldots)(1 + x^2 + x^4 + x^6 + \ldots)$$
$$(1 + x^3 + x^6 + \ldots)(1 + x^4 + \ldots)(1 + x^5 + \ldots)(1 + x^6 + \ldots) \cdot \ldots$$

Multiplying and simplifying carefully, we get

$$P(x) = 1 + x + 2x^2 + 3x^3 + 5x^4 + 7x^5 + 11x^6 + \ldots$$

Oftentimes we are more interested in special subclasses of partitions. Generating functions prove useful in describing and understanding them as well. We begin with two definitions: Let $p_u(n)$ be the number of partitions of n into unequal parts and let $p(n, o)$ be the number of partitions of n containing odd parts only. For example, the partitions of 5 into unequal parts are $5, 4 + 1$, and $3 + 2$. All other partitions of 5 include some repetition. Thus $p_u(5) = 3$. The partitions of 5 containing odd parts only are $5, 3 + 1 + 1$, and $1 + 1 + 1 + 1 + 1$. Hence $p(5, o) = 3$ as well. Euler was the first to prove that in fact, for any n, $p_u(n)$ equals $p(n, o)$. I think you'll agree that the result is hardly obvious.

It is an easy matter to determine the generating functions for each of these special partition functions. Again, it is convenient to define $p_u(0)$ and $p(0, o)$ to be one. The justification for Proposition 15.4 is saved as an exercise. But please multiply out the first few factors of each generating function in order to fully understand what's going on.

Proposition 15.4:

$$(a) \prod_{k=1}^{\infty} (1 + x^k) = \sum_{n=0}^{\infty} p_u(n) x^n.$$

$$(b) \prod_{k=1}^{\infty} \frac{1}{1 - x^{2k-1}} = \sum_{n=0}^{\infty} p(n, o) x^n.$$

Proposition 15.5: *For all* $n \geq 0, p_u(n) = p(n, o)$.

We will provide two proofs of Proposition 15.5. The first one is really slick and quite elegant. Even so, one might feel unsatisfied not really understanding why there is a one-to-one correspondence between partitions with unequal parts and partitions consisting solely of odd parts. The second proof is more direct in that it sets up such a correspondence between the two types of partitions. In any event, both proofs are very instructive and a joy to read and absorb.

First Proof of Proposition 15.5 (L. Euler): *By Proposition 15.4, it suffices to show that*

$$\prod_{k=1}^{\infty}(1+x^k) = \prod_{k=1}^{\infty}\frac{1}{1-x^{2k-1}}.$$

But $(1+x^k)(1-x^k) = 1 - x^{2k}$. *Hence*

$$\prod_{k=1}^{\infty}(1+x^k) = \prod_{k=1}^{\infty}\frac{1-x^{2k}}{1-x^k}.$$

On the right-hand side, the numerator contains all the even exponents, while the denominator contains all exponents, odd and even alike. Hence all the even exponent factors divide out. Thus

$$\prod_{k=1}^{\infty}\frac{1-x^{2k}}{1-x^k} = \prod_{k=1}^{\infty}\frac{1}{1-x^{2k-1}}.$$

The result follows. □

THE Sylvester Medal, to be awarded by the ROYAL SOCIETY triennially for the encouragement of Mathematical research. This reproduction is presented to the subscribers in memory of the late Professor J. J. SYLVESTER, by Professor J. M. Peirce of Cambridge, Massachusetts, U.S.A.

July, 1899.

The second proof we offer is due to the immensely versatile and talented mathematician James Joseph Sylvester (1814–1897). Sylvester was born in London and studied at Cambridge University, earning the rank of Second Wrangler. However, being Jewish, he was barred from receiving a degree. Instead he earned a degree at Trinity College in Dublin. Sylvester was a very prolific scholar and held teaching posts at the Royal Military Academy at Woolrich as well as at the University of Virginia and Johns Hopkins University. Sylvester contributed several fundamental ideas in algebra and number theory—even coining the word *matrix*, now ubiquitous even in the movies! Sylvester also delighted in sprinkling apt quotations from ancient Greek and Latin literature within his papers. Though as well known for his eccentricities and absentmindedness, Sylvester encouraged and helped mold a new generation of mathematicians and was the principle force behind the emergence of an American mathematical community. He made several contributions to our understanding of the partition function, publishing his final result in this area in 1896 at the age of 82.

Second Proof of Theorem 15.5 (J.J. Sylvester): *On the one hand, let p be a partition of n into unequal parts. For each part r of the partition p, we can express r uniquely as $2^k m$ where m is odd. Now rewrite r as a sum of 2^k copies of m. By then adding all the rewritten parts of p in this manner, we obtain a partition p' of n having odd parts only. Two different partitions of n into unequal parts cannot be transformed to the same partition with odd parts only. If so there would have to be two different sums of distinct powers of 2 being equal. But if we write the numbers in binary, each power of 2 consists of a single one followed by all zeros. When we then add distinct powers of 2, there are no carry overs, and thus the two sums could not be equal.*

On the other hand, let p' be a partition of n into odd parts only. Suppose p' contains odd part m a total of k times (and perhaps other parts of other odd sizes). Write k uniquely as a sum of distinct powers of 2. Then create new parts by multiplying m by each such power of 2. Do the same for all odd parts of p'. In this way, we create a new partition p of n consisting of unequal parts. By the Fundamental Theorem of Arithmetic, $2^a m_1 \neq 2^b m_2$ for distinct odd m_1 and m_2. Thus no two partitions of n with odd parts only can be transformed into the same partition of n with unequal parts.

Therefore, for any given n we have constructed a one-to-one correspondence between the partitions of n with odd parts only and partitions of n with unequal parts. The result follows. □

An example to illustrate our second proof may be helpful. For example, consider the partitions of $n = 6$ into unequal parts. They are $6, 5 + 1, 4 + 2$, and $3+2+1$. The partition 6 can be rewritten as $2 \cdot 3$. As such 6 gets transformed to $3+3$, a partition with odd parts only. Similarly, 5 and 1 are rewritten as $1 \cdot 5$ and $1 \cdot 1$, and thus $5+1$ becomes itself, $5+1$. The partition $4+2$ is $4 \cdot 1+2 \cdot 1$, and so becomes $1+1+1+1+1+1$. Finally, $3+2+1 = 1 \cdot 3+2 \cdot 1+1 \cdot 1$, and thus transforms to $3+1+1+1$.

Here's an example of how the construction works in the opposite direction. Consider the partitions of $n = 5$ into odd parts only. They are $5, 3+1+1$, and $1+1+1+1+1$. The partition 5 contains only one copy of the odd part 5. Since $5 \cdot 1 = 5$, the partition 5 gets transformed to itself. Next consider $3+1+1$. The number 3 remains as 3, but $1+1$ becomes 2. Hence $3+1+1$ becomes $3+2$. Finally, $1+1+1+1+1$ consists of five ones. Since $5 = 4+1$, we rewrite $1+1+1+1+1$ as $4+1$. To check your understanding, you might wish to reverse the process in our two examples.

Now we introduce another pair of functions based on two disjoint but distinguished subclasses of partitions. We have already discussed $p_u(n)$, the number of partitions of n into unequal parts. Define the *even partition function* $E(n)$ to be the number of partitions of n into an even number of unequal parts, and define

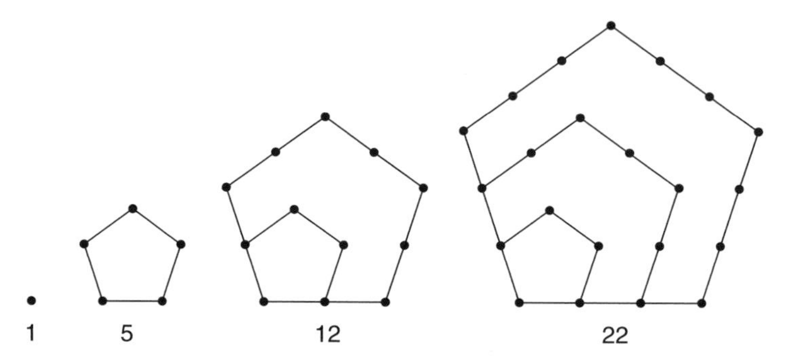

Figure 15.1 Pentagonal numbers.

the *odd partition function* $O(n)$ to be the number of partitions of n into an odd number of unequal parts. For example, $E(5) = 2$ since the two appropriate partitions are $3 + 2$ and $4 + 1$. Analogously, $O(5) = 1$ since 5 is the only partition of 5 into an odd number of unequal parts. Similarly, $E(6) = 2$ corresponding to the partitions $4 + 2$ and $5 + 1$, and $O(6) = 2$ to account for the partitions 6 and $3 + 2 + 1$. Necessarily, $E(n) + O(n) = p_u(n)$ for all $n \geq 1$.

Next we turn our attention to a wonderful result of which tells us when $E(n)$ and $O(n)$ are equal, and if not, when and by how much do they differ. Miraculously it happens that the functions $E(n)$ and $O(n)$ are related to a set of numbers called pentagonal numbers, so named because of their relation to simple five-sided geometric designs. Furthermore, our result will lead immediately to a fairly useful way to extend our calculations of $p(n)$. We begin by recalling the pentagonal numbers $g(n) = 1 + 4 + 7 + \ldots + (3n - 2)$ for any $n \geq 1$. The nth pentagonal number can be graphically displayed as a group of dots arranged in a pentagon (Fig. 15.1).

By induction it can readily be shown that $g(n) = \frac{n(3n-1)}{2}$ for all $n \geq 1$. Hence the sequence of pentagonal numbers begins 1, 5, 12, 22, 35, etc. We now extend the definition of pentagonal numbers to include negative values of the argument as well. The numbers $g(-n)$ for $n \geq 1$ no longer correspond to a five-sided group of dots, but thus is the nature of mathematical generalization. Just go with the flow. The additional numbers are $g(-n) = \frac{-n(-3n-1)}{2} = \frac{n(3n+1)}{2}$. These yield the numbers 2, 7, 15, 26, 40, etc. There is no overlap between the new pentagonal numbers with negative argument and those with positive. The reason is that for each n, $g(n) < g(n) + n = \frac{n(3n-1)}{2} + n = \frac{n(3n-1)+2n}{2} = \frac{n(3n+1)}{2} = g(-n) < g(n) + (3n - 2) = g(n + 1)$. So our full list of pentagonal numbers becomes 1, 2, 5, 7, 12, 15, 22, 26, 35, 40, etc.

So how do $E(n)$ and $O(n)$ relate to pentagonal numbers? And who would ever notice such a thing anyway? Drum roll, please ....

Theorem 15.6 (Pentagonal Number Theorem): *For all* $n \geq 1$, $E(n) = O(n)$ *unless* $n = \frac{k(3k \pm 1)}{2}$ *is a pentagonal number, in which case* $E(n) + (-1)^k = O(n)$.

Leonhard Euler discovered the Pentagonal Number Theorem after carrying out extensive calculations. The Pentagonal Number Theorem was stated and proved by him in 1742. The proof was intricate. We will end the chapter with a proof dated 1881 due to F. Franklin, a student of J.J. Sylvester's, and later an influential professor in his own right at Johns Hopkins University. But before we get to the proof, let us see how Euler made good use of this result.

Consider the difference function defined by $d(0) = 1$ and $d(n) = E(n) - O(n)$ for $n \geq 1$. Notice that $d(n)$ is always -1, 0, or 1 by the Pentagonal Number Theorem. In fact, we can make a brief table of values of $d(n)$:

n	0	1	2	3	4	5	6	7	8	9	10	11	12	13	14	15	16	17
$d(n)$	1	-1	-1	0	0	1	0	1	0	0	0	0	-1	0	0	-1	0	0

Recall that the function $p_u(n) = E(n)+O(n)$ had generating function $\prod_{k=1}^{\infty}(1+x^k)$ since each exponent in the product appears exactly once, corresponding to each partition having unequal parts. If we tweak things slightly, we can obtain the generating function for $d(n) = E(n) - O(n)$. In particular,

$$\prod_{k=1}^{\infty}(1 - x^k) = \sum_{n=0}^{\infty} d(n)x^n.$$

All even partitions contribute positively while all odd partitions contribute negatively. Next notice that the generating function for $d(n)$ is the reciprocal of the generating function for the partition function, $p(n)$. So it follows that

$$\left(\sum_{n=0}^{\infty} d(n)x^n\right)\left(\sum_{n=0}^{\infty} p(n)x^n\right) = 1.$$

By equating corresponding coefficients of x^n, we get $d(0)p(0) = 1$ and

$$\sum_{k=0}^{n} d(k)p(n-k) = 0 \text{ for } n \geq 1. \tag{15.4}$$

By the Pentagonal Number Theorem, most of the terms in the sum in Formula 15.4 are zero. This makes it especially useful in calculating a particular partition number based on those with smaller argument. In particular, Formula 15.4 can be rewritten to give us

$$p(n) = p(n-1) + p(n-2) - p(n-5) - p(n-7) + \ldots + (-1)^{k+1}$$
$$\left\{p(n - \frac{k(3k-1)}{2}) + p(n + \frac{k(3k+1)}{2})\right\} + \ldots$$

For example, suppose we know the values of $p(n)$ for $n \leq 6$ and wish to compute $p(7)$. We have that $p(7) = p(6)+p(5)-p(2)-p(0) = 11+7-2-1 = 15$. This is a lot less work than that of listing all 15 partitions of 7. Next we compute the eighth partition number. By Euler's Identity, $p(8) = p(7)+p(6)-p(3)-p(1) = 15+11-3-1 = 22$. In this way, we could extend our list of partition numbers indefinitely.

Proof of Theorem 15.6: *Let $n \geq 1$ and let p be a partition of n into unequal parts. Let s be the smallest part of p and let $t_1 > t_2 > \ldots$ be the largest parts in descending order. Let k be the maximum number for which $t_1 = t_2 + 1 = \ldots = t_k + (k-1)$. Thus k is the length of the string of numerically consecutive parts in p beginning with the largest part. There are two possibilities: (i) $s \leq k$, or (ii) $s > k$.*

(i) If $s \leq k$, then transform p to another partition p' of n by removing the part s and adding one to each of the parts $t_1, \ldots, t_s$. By our construction, the partition p' is also a partition of n into unequal parts. Furthermore, p' has one fewer part than p. Hence they are of opposite parity, namely one has an even number of parts and other an odd number of parts. Thus the two partitions contribute equally to the totals $E(n)$ and $O(n)$, maintaining equity between them. Note further that if we define s' and k' analogously for the partition p', then $s' > k'$. So if p is of type (i), then p' is of type (ii).

There is a possible exceptional case where our construction from a type (i) to a type (ii) partition cannot be completed. In the situation where both $s = k$ and $s = t_k$, when we remove s there are no longer k consecutive large parts to which we can add 1. Namely, part t_k is missing. In this situation, we know the exact form of p. It is the partition $n = (2k-1) + (2k-2) + \ldots + k$ consisting of k consecutive integers beginning with k. But we can readily simplify to obtain $n = [1 + \ldots + (2k-1)] - [1 + \ldots + (k-1)] = \frac{(2k-1)2k}{2} - \frac{(k-1)k}{2} = \frac{k(3k-1)}{2}$, a pentagonal number.

(ii) If $s > k$, then transform p to p' by subtracting 1 from each of $t_1, \ldots, t_k$ and adding an additional part k. Notice that the new part k is necessarily the smallest part of p' since s, the smallest part of p, satisfied $s < k$. So p' is a partition of n into unequal parts. The number of parts of p and p' differ by one. So one is an even partition, the other one odd. Furthermore, by our construction p' satisfies $s \leq k$.

Again there is a possible exceptional case where our construction fails. Suppose that $s = k + 1$ and $s = t_k$. In this case, the permutation p' would have two equal smallest parts $s - 1$ and hence not be a partition with unequal parts. In this situation n is the pentagonal number $2k + (2k-1) + \ldots + (k+1) = \frac{k(3k-1)}{2} + k = \frac{k(3k+1)}{2}$.

Since $(p')' = p$ in all but the exceptional cases, barring them there is a one-to-one correspondence between unequal partitions of n into an even or odd number of parts, respectively. In the two exceptional cases, k is the number of parts of p. Thus in the case when n is the pentagonal number $\frac{k(3k\pm1)}{2}$, there is an additional odd permutation when k is odd and there is an additional even permutation when k is even. No value of n has more than one exceptional case. Therefore, $E(n) = O(n)$ unless $n = \frac{k(3k\pm1)}{2}$, in which case $E(n) - O(n) = (-1)^k$. □

WORTH CONSIDERING

1. Show that the number of partitions of n into an even number of parts is equal to the number of partitions of n with largest part an even number.
2. Show that the number of partitions of n containing no part k is $p(n)-p(n-k)$ for any $k \leq n$.
3. **(a)** What is the generating function for the function $p(n, 3)$, the number of partitions of n with largest part at most 3?

   (b) Use the generating function to calculate $p(6, 3)$. Compare with computing $p(6, 3)$ directly.
4. **(a)** Let $p(n, e)$ be the number of partitions of n having even parts only. Compute $p(n, e)$ for $1 \leq n \leq 10$.

   (b) Determine the generating function for $p(n, e)$.

   (c) What is the relation between $p(n, e)$ and the standard partition function, $p(n)$?
5. Let $p_u(n, e)$ be the number of partitions of n into even and unequal parts, and let $p_u(n, o)$ be the number of partitions of n into odd and unequal parts. Determine the generating functions for both $p_u(n, e)$ and $p_u(n, o)$.
6. **(a)** Make a table for $p_u(n, o)$ for $1 \leq n \leq 10$.

   (b) Recall that a self-conjugate partition is one for which its conjugate is identical. Let $\mathrm{sc}(n)$ be the number of self-conjugate partitions of n. Make a table for $sc(n)$ for $1 \leq n \leq 10$. Discuss the result.
7. Let $c(n)$ be the number of compositions of n. Show that the number of compositions of n with no part 1 is F_{n-1} where F_n is the nth Fibonacci number.
8. Use Euler's Identity to calculate $p(9)$ and $p(10)$.
9. Twins can be either fraternal or identical. Triplets can be all identical, two identical and one fraternal, or all fraternal. How many possibilities are there for quadruplets? Quintuplets? Sextuplets? Septuplets?

16 Pascal Potpourri

Blaise Pascal

In this section we will present several interesting properties of what is commonly called Pascal's triangle, hence the title of this chapter. Blaise Pascal (1623–1662) was a great mathematical prodigy, despite his father's firm insistence that he devote himself only to the study of languages and the classics. As a youngster he discovered the theorem that the sum of the angles of any triangle is 180 degrees. He then went on to rediscover many of the first 32 propositions of Euclid's *Elements*. At this point, his father recanted his ban on Blaise's mathematical

Mathematical Journeys, by Peter D. Schumer
ISBN 0-471-22066-3

studies and presented him with a copy of the great geometric work. By the age of 12, the young Pascal had mastered it completely and was making his own discoveries to boot. By age 14, he had begun attending the weekly gatherings of the French mathematicians and was discussing and debating with the likes of René Descartes, Marin Mersenne, and G.P de Roberval. At age 16 he wrote a treatise *On Conics*, which added important new discoveries to a field thought to be so well trodden by the great Apollonius (ca. 250–175 B.C.E.) and other mathematicians from ancient times. Pascal's essay, which included his "mystic hexagram theorem," was a precursor to the development of projective geometry. He enjoyed practical invention and designed something still widely used, namely the one-wheeled wheelbarrow. At age 18 he invented a mechanical calculating machine to help with his father's bookkeeping. The invention was a success and Pascal built and sold over 50 such machines. (It is for this reason that the modern computer language Pascal was named in his honor.) Pascal also made some interesting discoveries about fluid action under air pressure and studied many geometric shapes, including the Archimedean spiral.

But Pascal was tormented by an unremitting religious fervor and also by ill health, chronic insomnia, and dyspepsia. Unfortunately for the development of mathematics, a great deal of his time was devoted to religious contemplation and masochistic self-injury. He wished "to contemplate the greatness and the misery of man," as he put it. His well-deserved acclaim as a great French philosopher and writer is due less to his mathematics and more to his *Provincial Letters*, directed against the Jesuits, and his *Pensées*, written later in his brief life. He only returned to mathematics fitfully when he took personal events as religious signs. In 1658, while thinking about the cycloid during one of many sleepless nights, he noticed that his toothache completely went away. The cycloid is the curve sketched by a point on the circumference of a circle as it moves along a straight line. Interpreting this sudden burst of good health as divine intervention, he spent the next eight days on nothing but mathematical studies before returning to his other devotions.

Together with Fermat, Pascal helped develop the mathematical theory of probability via their famous correspondence concerning a gambling problem. In particular, the problem dealt with the fairest way to divide a pot of money to two contestants involved in an uncompleted game involving several rounds. The details of the problem are not important here, but the triangular array of numbers that Pascal introduced to solve the problem is relevant. The nth row of this numerical triangle consists of the number of ways to choose a set of k items from a set of n distinguishable objects for all k with $0 \leq k \leq n$. The top of the triangle is the zeroth row consisting of the number 1. Figure 16.1 shows how the triangle begins.

For example, the third row consists of the numbers 1, 3, 3, 1. If we have three marbles colored red, white, and blue, there is just one way to not choose any of them. Hence the first entry is the number 1. The second entry is 3, corresponding to the three ways to choose one marble. Similarly there are three ways to choose two marbles out of three, and finally just one way to choose all three marbles.

Pascal's triangle had actually been studied by many people long before Pascal himself rediscovered it. The triangular array of numbers and its use in

							1							
						1		1						
					1		2		1					
				1		3		3		1				
			1		4		6		4		1			
		1		5		10		10		5		1		
	1		6		15		20		15		6		1	
1		7		21		35		35		21		7		1

Figure 16.1 The arithmetic triangle.

combinatorial matters was certainly known by the medieval Jewish mathematician and astronomer, Levi Ben Gerson (1288–1344). But before him, in the 11th century, the Chinese mathematician Jia Xian used the triangle at least implicitly to further develop and generalize procedures for finding square and cube roots. Jia Xian's works have been lost, but they are referred to in extant work by Yang Hui dated about 1261. Certainly the use of the arithmetic triangle to expand polynomials was fully understood by then. For example, to expand $(x + y)^3$, the third row of the triangle provides the coefficients to $1x^3 + 3x^2y + 3xy^2 + 1y^3$. We formalize this observation below as the binomial theorem, but its basic idea must go back to the Chinese. Or does it? There is reliable documentation that the Persian scholar Omar Khayyam (ca. 1044–1123) used the arithmetical triangle. Even more distant, the Islamic scientist, Ibn al-Haytham (965–1039), better known in the West as Alhazen, seems to have had some facility with the triangle at least for small values. His formulas for summing cubes and fourth powers relied on it. And there seems to be some similar work done by Abu Bakr al-Karaji (d. 1019). The point of all of this discussion is that the study of the origins of great ideas is never ending. The deeper mathematics historians dig, the more treasures they discover.

The term *Pascal's triangle* is certainly inappropriate if we want to give credit to the first person to *discover* it. So why does the arithmetic triangle continue to be named after Pascal? Certainly a lack of knowledge of ancient cultures, ancient languages, and an ignorance of the histories of non-Western peoples together with a Eurocentric mind set are largely to blame. But Pascal's use of the triangle was completely explicit and he did make several important discoveries with it. So perhaps honoring him by name is not wholly inappropriate as long as we are fully cognizant that his truly substantial contributions are but a small part of the fuller story.

Recall the binomial coefficient $\binom{n}{k} = \frac{n!}{k!(n-k)!}$ defined for $n \geq 1$ and $0 \leq k \leq n$. The binomial coefficient $\binom{n}{k}$ counts the number of different ways that k items can be chosen from a set of n distinguishable objects. The term *binomial coefficient* comes from its use in the following key result:

Binomial Theorem: *For $n \geq 1$, $(x + y)^n = \sum_{k=0}^{n} \binom{n}{k} x^{n-k} y^k$.*

The theorem is fairly intuitive. In multiplying $(x + y)$ times itself n times, the sum of the exponents for x and y must be n. In addition, for each k the coefficient of $x^{n-k}y^k$ derives from choosing k factors of y out of n possible choices. There are precisely $\binom{n}{k}$ such choices. But then there is just one way to choose the remaining x's.

For example, $(x + y)^4 = \binom{4}{0}x^4 + \binom{4}{1}x^3y + \binom{4}{2}x^2y^2 + \binom{4}{3}xy^3 + \binom{4}{4}y^4 = x^4 + 4x^3y + 6x^2y^2 + 4xy^3 + y^4$. Hence the coefficients can be read off the appropriate row of Pascal's triangle. Furthermore, with proper choice of x and y we can make some interesting discoveries.

Proposition 16.1: *For $n \geq 1$ and $m \varepsilon \mathbf{Z}$,*

(a) $\sum_{k=0}^{n} \binom{n}{k} = 2^n$
(b) $\sum_{k=0}^{n} (-1)^k \binom{n}{k} = 0$
(c) $\sum_{k=0}^{n} \binom{n}{k} m^k = (m + 1)^k$.

Proof of Proposition 16.1:

(a) In the binomial theorem, let $x = 1$ and $y = 1$.
(b) Let $x = 1$ and $y = -1$.
(c) Let $x = 1$ and $y = m$. □

Result 16.1 (a) can be interpreted as expressing the fact that the total number of subsets of a set of n objects is 2^n as long as we include the empty set and the set itself as legitimate subsets. (Of course, another way to see that there are 2^n subsets is by noting that for each of the n elements, we must make a yes or no decision when deciding whether or not to include the item. Hence there are $2 \cdot 2 \cdot \ldots \cdot 2 = 2^n$ possible decisions, one for each possible subset.) Part (b) says that the alternating sum across any row of Pascal's triangle is zero. Equivalently, in any row, the sum of the entries in the odd-numbered positions equals the sum of the entries in the even-numbered positions. Result (c) is an extension of part (b) in the sense that (b) is the special case for $m = -1$.

The combinatorial aspects of Pascal's triangle is further exposed by noting that each interior entry (i.e., any entry not a 1) is the sum of the two entries just above it. For example, 35 in the seventh row of the triangle is the sum of 15 and 20. In general, this is how additional rows of Pascal's triangle are easily generated. To verify this, we need to show that $\binom{n}{k} + \binom{n}{k+1} = \binom{n+1}{k+1}$ for any $k < n$. We can verify this directly via algebraic manipulation:

$$\binom{n}{k} + \binom{n}{k+1} = \frac{n!}{k!(n-k)!} + \frac{n!}{(k+1)!(n-k-1)!}$$
$$= \frac{(k+1)n! + (n-k)n!}{(k+1)!(n-k)!}$$

$$= \frac{(n+1)n!}{(k+1)!(n-k)!}$$
$$= \binom{n+1}{k+1}.$$

Alternatively, we gain additional insight by arguing from a combinatorial perspective. There are $\binom{n+1}{k+1}$ subsets of the set of numbers $S = \{1, 2, \ldots, n+1\}$ which contain $k+1$ elements. Some of these subsets contain the number 1 and some do not. The number of such subsets of S that do not contain the number 1 is $\binom{n}{k+1}$ since we must choose $k+1$ numbers from among the n numbers $2, \ldots, n+1$. The number of such subsets of S that do contain the number 1 is $\binom{n}{k}$ since once we include the number 1, then we must choose an additional k numbers from among $2, \ldots, n+1$. These comprise all the $k+1$ element subsets of S, and hence the result follows.

Here is a cute result due to an observation of Paul Erdös, published in 1972. No doubt, others have noticed the same property, but I am not aware of an earlier reference.

Proposition 16.2: *Excluding the two 1's, any two entries in the same row of Pascal's triangle are not relatively prime.*

Proof of Proposition 16.2: *Let $0 < i < j < n$. We will show that $\gcd\left(\binom{n}{i}, \binom{n}{j}\right) > 1$. We begin by multiplying $\binom{n}{j}$ by $\binom{j}{i}$.*

$$\binom{n}{j}\binom{j}{i} = \frac{n!}{j!(n-j)!}\frac{j!}{i!(j-i)!}$$

and upon rearranging and multiplying numerator and denominator by $(n-i)!$

$$= \frac{n!}{i!(n-i)!}\frac{(n-i)!}{(n-j)!(j-i)!}$$
$$= \binom{n}{i}\binom{n-i}{j-i}.$$

If the $\gcd\left(\binom{n}{i}, \binom{n}{j}\right) = 1$, then $\binom{n}{i} | \binom{n}{j}$ since $\binom{n}{i} | \binom{n}{j}\binom{j}{i}$ as shown above. But

$$\binom{n}{i} = \frac{n!}{i!(n-i)!}$$
$$= \frac{n(n-1)\cdots(n-i+1)}{i!}$$
$$> \frac{j(j-1)\cdots(j-i+1)}{i!}$$

$$= \frac{j!}{i!(j-i)!}$$

$$= \binom{j}{i}.$$

This contradicts the assertion that $\binom{n}{i} | \binom{j}{i}$, *since a larger natural number cannot divide evenly into a smaller one. Hence* $gcd\left(\binom{n}{i}, \binom{n}{j}\right) > 1$. □

Next we make some interesting, but fairly transparent observations about Pascal's triangle. If we look down the first diagonal column starting at the top and moving down and to the left, all entries are 1. The second diagonal column contains all the counting numbers in order. The third diagonal column contains all the triangular numbers 1, 3, 6, 10, etc. The reason is that for $k \geq 2$, the kth entry is $\binom{k}{2} = \frac{k(k-1)}{2} = 1 + 2 + \ldots + (k-1)$. The fourth diagonal column contains the *tetrahedral* numbers 1, 4, 10, 20, 35, etc., which can be viewed as three-dimensional analogs of the triangular numbers (Fig. 16.2).

Here's a nice little combinatorial result.

Proposition 16.3: *For* $n \geq 2m - 1$, *the number of* m*-element subsets of the set* $S = \{1, 2, \ldots, n\}$ *not containing two consecutive numbers is* $\binom{n-m+1}{m}$.

Proof of Proposition 16.3: *Consider* n *markers of which* m *are purple and* $n-m$ *are brown. Set aside* $m - 1$ *brown markers. This is possible since* $m - 1 = (2m-1)-m \leq n-m$. *Now place the remaining* $n-(m-1) = n-m+1$ *markers arbitrarily from left to right. There are* $\binom{n-m+1}{m}$ *ways to arrange them since* m *are purple and the rest are brown. Finally, intersperse the remaining* $m-1$ *brown markers between the* m *purple ones. There is only one distinguishable way to do this. So there are* $\binom{n-m+1}{m}$ *ways to place the* n *markers with* m *purple, the rest brown, and no two purple markers adjoining one another. Now think of the purple*

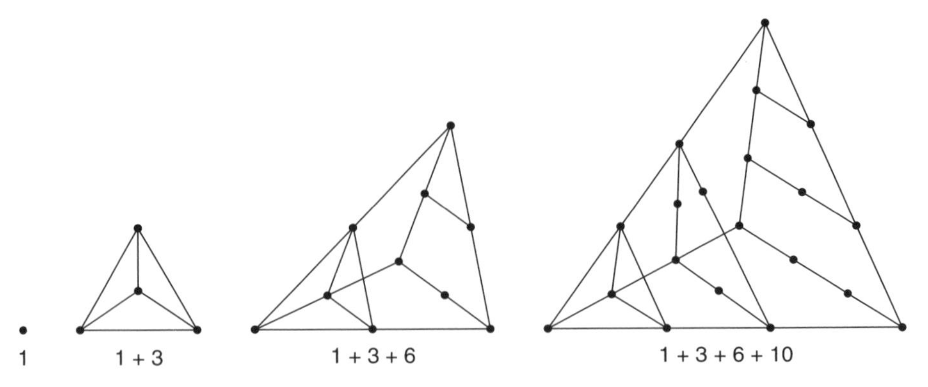

Figure 16.2 Tetrahedral numbers.

markers as representing chosen numbers from S and brown markers as numbers not chosen. There is a one-to-one correspondence between m-element subsets of S with no two consecutive numbers and arrays of m purple and n − m brown markers with no two purple markers juxtaposed. The result follows immediately.□

Now we combine Pascal's triangle with the Fibonacci numbers!

Proposition 16.4: *The total number of nonempty subsets of the set* $S = \{1, 2, \ldots, n\}$ *not containing two consecutive numbers is* $F_{n+2} - 1$.

Proof of Proposition 16.4 (Induction on *n*): *If* $n = 1$, *then there is just one such subset of S, namely S itself. Since* $1 = F_3 - 1$, *the proposition holds in this case. Next assume the result is true for all sets up to order* $n-1$. *For* $S = \{1, 2, \ldots, n\}$, *consider all nonempty subsets not containing two consecutive numbers. Let us separate these subsets into those that contain the number n and those that do not. If a subset contains the number n, then it cannot contain the number* $n-1$. *Hence the number of subsets of S containing n and not having consecutive members is in one-to-one correspondence with the subsets of* $\{1, 2, \ldots, n-2\}$ *having no consecutive numbers, but with the empty set included (corresponding to the subset* $\{n\}$ *of S). By our inductive assumption, the number of such nonempty sets was* $F_n - 1$. *Thus the number of nonempty subsets of S containing n with no consecutive numbers is* F_n. *The nonempty subsets of S not containing n with no consecutive numbers is identical to the nonempty subsets of* $\{1, 2, \ldots, n-1\}$ *with no consecutive elements. By our inductive assumption, the number of such subsets is* $F_{n+1} - 1$. *Therefore, the total number of nonempty subsets of S is* $F_n + F_{n+1} - 1 = F_{n+2} - 1$. □

Combining Propositions 16.3 with Proposition 16.4, we have the following result. Note that $[x]$ represents the greatest integer less than or equal to x.

$$\sum_{m=1}^{[\frac{n+1}{2}]} \binom{n-m+1}{m} = F_{n+2} - 1. \tag{16.1}$$

We can interpret this as a sum of appropriate entries in Pascal's triangle. For example, for $n = 7$, we have $\binom{7}{1} + \binom{6}{2} + \binom{5}{3} + \binom{4}{4} = 33 = F_9 - 1$. Graphically, we are adding the bold-faced entries in Figure 16.3.

Now we move on to a fun problem concerning food preparation at a summer camp!

Problem: *A summer camp has 250 campers. For every pair of campers, A and B, there is an item of food that A is willing to eat but B is not and vice versa. What is the smallest total number of food items that could conceivably be made available at mealtime so that every camper will eat something?*

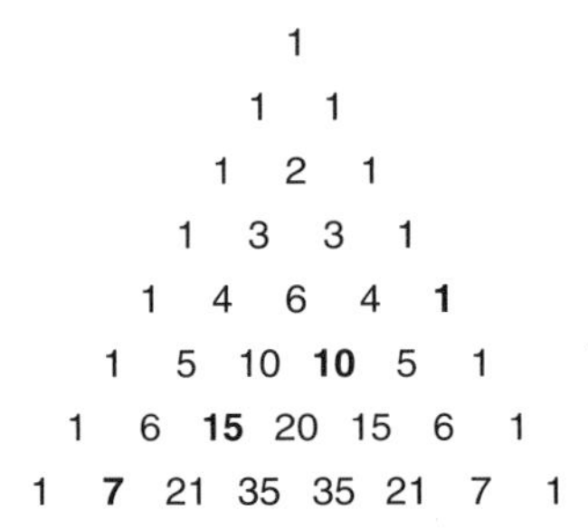

Figure 16.3 Fibonacci numbers occur as sums within the arithmetic triangle.

Certainly with really rotten luck, the kitchen staff might have to purchase 250 different types of food if there were no overlap in tastes among any of the campers. But here we want the best possible scenario under the given condition.

At first, this problem may seem to be completely unrelated to all that we have discussed so far. But binomial coefficients provides the key. If we continue Pascal's triangle through row 10, the central number in the bottom row is the number $\binom{10}{5} = 252$. Hence there are 252 ways to choose 5 food items out of a set of 10. So if each of the 250 campers likes a different subset of 5 food choices, then our conditions are met. Hence the answer is a measly 10.

While we're in the neighborhood, there are some interesting questions about these central binomial coefficients. In 1978, Erdös conjectured that the binomial coefficient $\binom{2n}{n}$ was never squarefree for $n > 4$. The conjecture was correct and was established for n sufficiently large by A. Sárközy in 1985 and in general by A. Granville and O. Ramaré in 1993. In 1975, P. Erdös, R.L. Graham, I.Z. Ruzsa, and E. Straus proved that for any two primes p and q, there are infinitely many values of n for which $\gcd(\binom{2n}{n}, pq) = 1$. However, so far no one can prove a similar result for three primes, in fact for *any* three primes. In this regard, Ron Graham has offered \$100 to anyone who can show that $\gcd(\binom{2n}{n}, 105) = 1$ infinitely often.

Our final result is a stunning primality test due to Daniel Shanks (1917–1996) and H.B. Mann. A few words about the first author might be in order here. Shanks worked as a physicist at both the Aberdeen Proving Grounds and later at the Naval Ordnance Laboratory (NOL). While there he wrote his Ph.D. dissertation despite not having attended any math graduate school. Since no university seemed prepared to grant a degree to a nonmatriculant, he didn't receive his graduate degree for several years, until he took the requisite courses at University of Maryland. By this point he headed the Applied Mathematics Laboratory at the NOL. Subsequent to that he had a long stint as Senior Research Scientist at the Naval Ship R&D Center at the David Taylor Model Basin, followed by a distinguished career as an Adjunct Professor to the Mathematics Department at University of Maryland, College Park. Shanks's main work dealt in numerical analysis, number theoretic problems, and particularly in computational methods in number theory. In all he published over 80 papers, many of which were highly influential regarding such areas of inquiry as calculating π, searching for primes of the form $n^2 + 1$, factoring large integers, primality testing, and calculating number theoretic invariants such as the class number of algebraic number fields. He named his algorithms with suggestive six-letter titles such as CLASNO, SQUFOF, REGULA, WHEEL8, EPZETA, SUMJAC, and EUPROD. He also had quite a sharp wit. At a number theory conference held at Howard University in 1979, one of the speakers gave a very nice lecture on the proof of one of his recent results. As soon as the lecture ended, someone in the audience stood up and explained that in fact he, the audience member, had already proved the same result. The paper was now being refereed and he didn't want anyone at the conference to think that he had stolen the result from the current speaker at the conference. At this point, Dan Shanks begrudgingly stood up and said, "While we're at it perhaps I should mention that this very result is in my 1962 book, *Solved and Unsolved Problems in Number Theory*," and he proceeded to cite the page and theorem number. Through his writing, teaching, and lifetime dedication to mathematics, he inspired many a budding mathematician, including the author of this book.

The primality test that we consider here makes direct use of Pascal's triangle. Actually, we need to distort the triangle a bit by sliding each row to the right two places farther than the row above. We then circle some numbers based on whether they are divisible by their row numbers, and then voilà, we have a simple necessary and sufficient primality test for the column numbers. Appropriately, the primality algorithm was called SHMAPT by Shanks—a nifty name for a nifty primality test. Here are the details.

Theorem 16.5 (Shanks-Mann Primality Test, 1972): *Make a rectangular table of values beginning with row and column zero in which the entries in the nth row consist of the binomial coefficients* $\binom{n}{j}$ *for* $0 \le j \le n$ *placed in column positions* $2n + j$. *Circle all the entries divisible by their row number. Then the column number is prime if and only if all entries under it are circled.*

Here is how the table of values begins in the Shanks-Mann Primality Test (Fig. 16.4).

Proof of Theorem 16.5: *Figure 16.4 verifies the claim for columns* $k = 1, 2,$ *and 3 (in fact, up to* $k = 17$*). Assume in what follows that* $k > 3$. *Either k is even or k is odd.*

***k* even:** *If* $k = 2m$ *with* $m > 1$, *the first entry of row m in column* $k = 2m$ *is* $\binom{m}{0} = 1$. *Hence it is not circled. So all even numbers greater than 2 are determined to be composite.*

***k* odd:** *We will show that if k is prime, then all entries in the* k^{th} *column are circled. Furthermore, if k is composite, then there is at least one uncircled entry in column* k.

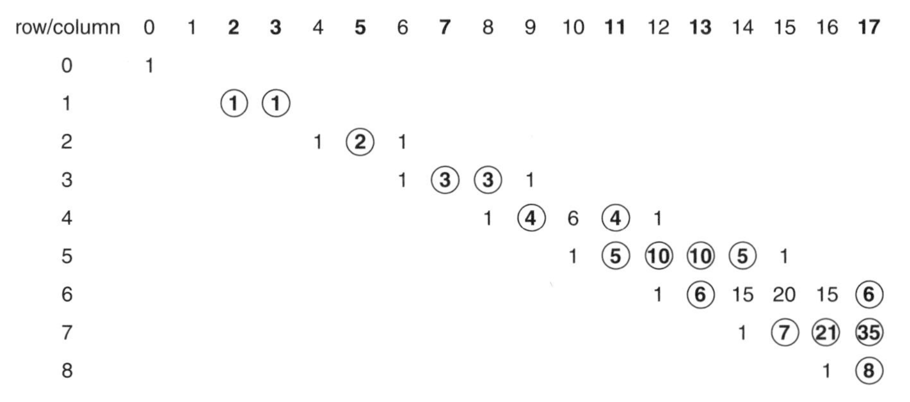

row/column	0	1	**2**	**3**	4	**5**	6	**7**	8	9	10	**11**	12	**13**	14	15	16	**17**
0	1																	
1			①	①														
2					1	②	1											
3							1	③	③	1								
4									1	④	6	④	1					
5											1	⑤	⑩	⑩	⑤	1		
6													1	⑥	15	20	15	⑥
7															1	⑦	㉑	㉟
8																	1	⑧

Figure 16.4 Shanks-Mann primality test.

The rows n which put entries in column k are those for which $2n \leq k \leq 3n$, that is, $\frac{k}{3} \leq n \leq \frac{k}{2}$. In fact, the entry in row n and column k is $\binom{n}{k-2n}$. We distinguish two cases depending on whether the column k is prime or an odd composite:

k = prime p: *In this case the entries in column k are $\binom{n}{p-2n}$ for all n satisfying the inequalities $\frac{p}{3} \leq n \leq \frac{p}{2}$. Since $p > 3$, we have that $1 < n < p$. Hence $\gcd(n,p) = 1$ for all such n. In fact, $\gcd(n, p-2n) = 1$ as well. Next we rewrite the binomial coefficient $\binom{n}{p-2n}$.*

$$\begin{aligned}
\binom{n}{p-2n} &= \frac{n!}{(p-2n)!(n-[p-2n])!} \\
&= \frac{n!}{(p-2n)!(3n-p)!} \\
&= \frac{n}{p-2n}\,\frac{(n-1)!}{(p-2n-1)!(3n-p)!} \\
&= \frac{n}{p-2n}\binom{n-1}{p-2n-1}.
\end{aligned}$$

Thus

$$(p-2n)\binom{n}{p-2n} = n\binom{n-1}{p-2n-1}.$$

Hence n divides $(p-2n)\binom{n}{p-2n}$. But since $\gcd(n, p-2n) = 1$, in fact n divides $\binom{n}{p-2n}$. Therefore, every entry in the pth column is circled.

k = odd composite: *Let p be an odd prime divisor of k. So $k = p(2r+1)$ for some integer $r \geq 1$. Thus $p \leq pr$ and $2pr < k = 2pr + p \leq 3pr$. Hence row*

$n = pr$ contributes of column k and its entry there is $\binom{n}{k-2n} = \binom{pr}{2pr+p-2p} = \binom{pr}{p}$. We next show that $n = pr$ does not divide $\binom{pr}{p}$, resulting in an uncircled entry in column k as claimed. To see this we expand $\frac{1}{pr}\binom{pr}{p}$.

$$\frac{1}{pr}\binom{pr}{p} = \frac{(pr-1)\cdots(pr-p+1)}{p(p-1)\cdots 2\cdot 1}.$$

Note that p divides the denominator, but p does not divide the numerator. Hence $\frac{1}{pr}\binom{pr}{p}$ is not an integer and pr does not divide $\binom{pr}{p}$. □

Albert Einstein (1879–1955) once said, "How can it be that mathematics, being after all the product of human thought independent of experience, is so admirably adapted to the objects of reality?" How true and how wonderful that the natural, outer world somehow seems destined to follow a mathematical logic that we humans can in our limited way penetrate and appreciate. Einstein was echoing the thoughts of another brave and great thinker from several centuries earlier. Galileo Galilei (1564–1642) wrote in his *Opere il Saggiatore*, "Philosophy is written in that great book which ever lies before our gaze—I mean the universe—but we cannot understand if we do not first learn the language and grasp the symbols in which it is written. The book is written in the mathematical language, and the symbols are triangles, circles and other geometrical figures, without the help of which it is impossible to conceive a single word of it, and without which one wanders in vain through a dark labyrinth."

But the Shanks-Mann Primality Test is a piece of pure mathematics and even there is not especially useful or practical. But it is certainly a very pretty result which expands the mind, lifts the spirit, and reminds us that seeking truth in mathematics is one sure path to the very heart of our most inner worlds. As Carl Gustav Jacobi (1804–1851) wrote, "... the only goal of Science is the honor of the human spirit, and that as such, a question of number theory is worth a question concerning the system of the world."

WORTH CONSIDERING

1. How many ways are there to deal n playing cards to two people? (Assume that the players may each receive any number of cards including none at all.)
2. How many ways are there to deal n playing cards to m people?
3. Does any row of Pascal's triangle have three consecutive entries in the ratio 1:2:3?
4. (Green Chicken Contest, 1998): Six boxes are numbered 1 through 6. How many ways are there to put 20 identical balls into these boxes so that none is empty?

5. How many ways can you distribute 40 identical candy bars to 10 children if each child must get at least one candy bar?

6. Prove that n is prime if and only if all binomial coefficients $\binom{n}{k}$ for $1 \leq k \leq n-1$ are divisible by n.

7. (Green Chicken Contest, 1986): How many South-East paths are there spelling MATHEMATICS? (One example is given below.)

M	A	T	H	E	M
A	T	H	E	M	A
T	H	E	M	A	T
H	E	M	A	T	I
E	M	A	T	I	C
M	A	T	I	C	S

8. (F. Mariares, 1913): Show that $\sum_{k=1}^{n} (2k-1)^2 = \binom{2n+1}{3}$.

9. (Chu Chi-kie, 1303): Show that $\sum_{k=0}^{r} \binom{n}{k}\binom{m}{r-k} = \binom{n+m}{r}$ for $r \leq n+m$.

10. Show that $\sum_{k=0}^{n} \binom{n}{k}^2 = \binom{2n}{n}$.

APPENDIX Comments and Solutions to Problems Worth Considering

1.1: $2 \cdot 3 \cdot 5 \cdot 7 + 1 = 211$ is prime, $2 \cdot 3 \cdot 5 \cdot 7 \cdot 11 + 1 = 2{,}311$ is prime, but $2 \cdot 3 \cdot 5 \cdot 7 \cdot 11 \cdot 13 + 1 = 30{,}031 = 59 \cdot 509$.

1.2: $2^{11} - 1 = 2{,}087 = 23 \cdot 89$.

1.3: If a prime $p | P_A$ then $p \nmid P_B$ and so $p \nmid (P_A + P_B)$. Similarly, if a prime $p | P_B$ then $p \nmid P_A$ and so $p \nmid (P_A + P_B)$. Thus any prime dividing $P_A + P_B$ must be an additional prime.

1.4: $7! + 2 = 5{,}042, \ldots, 7! + 7 = 5{,}047$ are six consecutive composites. But so are 114, ..., 119. In fact, 114, ..., 126 are all composite.

1.5: **(a)** $t_{n-1} + t_n = \frac{(n-1)n}{2} + \frac{n(n+1)}{2} = n^2$.

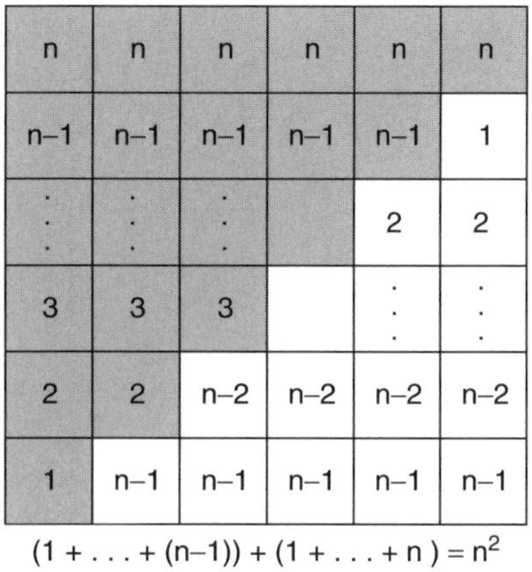

n	n	n	n	n	n
n−1	n−1	n−1	n−1	n−1	1
⋮	⋮	⋮		2	2
3	3	3		⋮	⋮
2	2	n−2	n−2	n−2	n−2
1	n−1	n−1	n−1	n−1	n−1

(1 + . . . + (n−1)) + (1 + . . . + n) = n^2

Figure A.1 $(1 + \ldots + (n-1)] + (1 + \ldots + n] = n^2$.

(b) $t_{n-1}^2 + t_n^2 = \frac{(n-1)^2 n^2}{4} + \frac{n^2(n+1)^2}{4} = \frac{n^2(n^2+1)}{2} = t_{n^2}$

Mathematical Journeys, by Peter D. Schumer
ISBN 0-471-22066-3

1.6: $\frac{n(3n-1)}{2} + (3n+1) = \frac{(n+1)(3n+2)}{2} = p_{n-1}$.

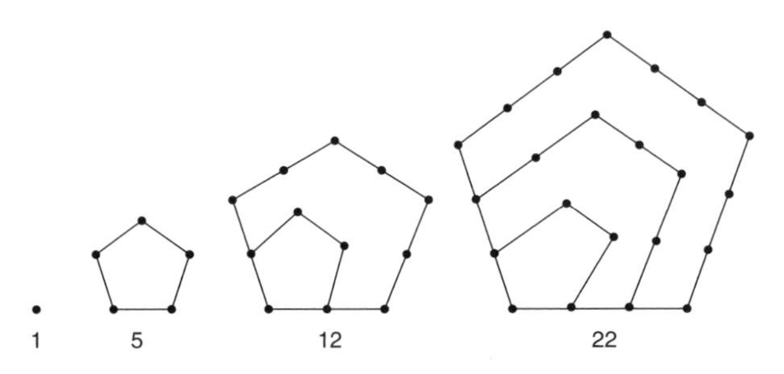

Figure A.2 Pentagonal numbers.

1.7: **(a)** Inductive step involves $(F_1 + \ldots + F_n) + F_{n+1} = (F_{n+2} - 1) + F_{n+1} = (F_{n+1} + F_{n+2}) - 1 = F_{n+3} - 1$.

(b) $(F_2 + \ldots + F_{2n}) + F_{2n+2} = (F_{2n+1} - 1) + F_{2n+2} = (F_{2n+1} + F_{2n+2}) - 1 = F_{2n+3} - 1$.

(c) $(F_1 + \ldots + F_{2n-1}) + F_{2n+1} = F_{2n} + F_{2n+1} = F_{2n+2}$.

1.8: Proof by induction is direct since $F_n F_{n+1} + F_{n+1}^2 = F_{n+1}(F_n + F_{n+1}) = F_{n+1} F_{n+2}$. In addition, the following diagram gives a geometric proof (A. Brousseau, *Fibonacci Quarterly*, 1972):

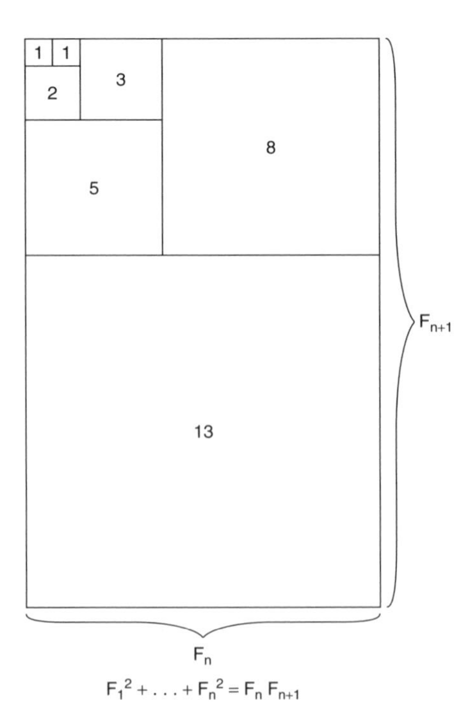

Figure A.3 $F_1^2 + \ldots + F_n^2 = F_n F_{n+1}$.

1.9: **(a)** Let $s(n)$ be the number of ways to place n stones in the bowl either one or two at a time. Note that $s(1) = 1 = F_2$ and $s(2) = 2 = F_3$. Generally, to place n stones in the bowl the first placement is either one stone or two stones. If we begin with one stone, then there are $s(n-1)$ ways to continue. If we begin with two stones, then there are $s(n-2)$ ways to continue. Thus $s(n) = s(n-1) + s(n-2)$, the same recurrence as the Fibonacci numbers.

(b) This problem is equivalent to part (a). Hence the answer is F_{n+1}.

1.10: Let $A = \begin{bmatrix} a_1 & b_1 \\ c_1 & d_1 \end{bmatrix}$ and $B = \begin{bmatrix} a_2 & b_2 \\ c_2 & d_2 \end{bmatrix}$. Then $\det(AB) = (a_1a_2 + b_1c_2)(c_1b_2 + d_1d_2) - (a_1b_2 + b_1d_2)(c_1a_2 + d_1c_2) = a_1a_2b_2c_1 + a_1a_2d_1d_2 + b_1b_2c_1c_2 + b_1c_2d_1d_2 - a_1a_2b_2c_1 - a_1b_2c_2d_1 - a_2b_1c_1d_2 - b_1c_2d_1d_2 = a_1a_2d_1d_2 - a_1b_2c_2d_1 - a_2b_1c_1d_2 + b_1b_2c_1c_2 = (a_1d_1 - b_1c_1)(a_2d_2 - b_2c_2) = \det A \cdot \det B$.

1.11: $19 = 13 + 5 + 1, 32 = 21 + 8 + 3, 232 = 144 + 55 + 21 + 8 + 3 + 1$.

1.12: L-shaped dominoes cannot cover a $3^n \times 3^n$ chessboard with or without one missing square. Since $4^n = 2^{2n}$, a $4^n \times 4^n$ chessboard with any single missing square can be covered with L-shaped dominoes.

1.13: The following figure shows two different decompositions of a square into ten smaller squares. There are 16 decompositions like the one in Figure A.4a and six like the one in Figure A.4b for a total of 22 decompositions.

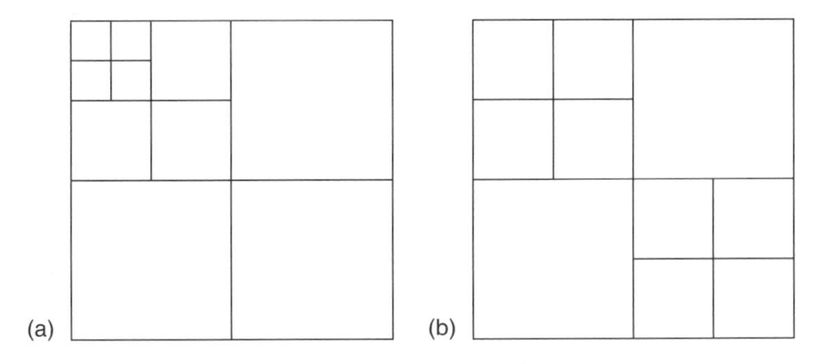

Figure A.4 Square decomposed into squares.

1.14: A $2 \times n$ chessboard can be covered with n 2×1 dominoes in F_{n+1} distinct ways.

1.15: The ten primes 2, 3, 5, 11, 23, 29, 41, 53, 83, and 89 are Germain primes.

1.16: $f(n) = n^2 - n + 41$ is prime for $n = 1, \ldots, 40$. But $f(41) = 41^2$.

2.1: Check initial case and then note that $\frac{(n+1)^3}{3} + \frac{(n+1)^2}{2} + \frac{n+1}{6} = \frac{n^3}{3} + \frac{n^2}{2} + \frac{n}{6} + (n^2 + 2n + 1)$.

2.2: Note that $n^5 - n = n(n-1)(n+1)(n^2+1)$. Now argue why $n^5 - n$ is divisible by 2, 3, and 5.

2.3: Mimic solution of Problem #1 modulo 10.

2.4: Twenty-three chocolate-covered strawberries cannot be ordered exactly, but all larger orders can be accomplished.

2.5: **(a)** Each handshake involves two people, so the total number must be even.

(b) If an odd number of people shook hands an odd number of times, then the total number of handshakes would be odd, contradicting part (a).

2.6: See Example #1 of Chapter 9.

2.7: In the first case, if all the British and Chinese history students talk to one another, then all 25 students could end up in American history. Similarly, in the second case, all 28 students could end up in American history. First two students must add British history and then the solution is the same.

2.8: This is the first theorem that I "discovered" and proved in college. Work modulo 3.

2.9: **(a)** The same proof that $\sqrt{2}$ is irrational works to show that $\sqrt{3}$ is irrational (or $\sqrt{n}$ for that matter where n is not a perfect square).

(b) In mimicking the proof of the irrationality of $\sqrt{2}$, we arrive at the step $4b^2 = a^2$. So $4|a^2$. But this *does not* imply that $4|a$.

(c) Show that $\sqrt{5}$ is irrational and then explain why a rational plus irrational is irrational and a rational times an irrational is irrational.

2.10: Since $2^7 = 128$ is approximately equal to $5^3 = 125$, it follows that 2^{10} is approximately equal to 10^3. Hence $\log_{10} 2$ is approximately equal to $\frac{3}{10} = 0.3$. (In fact, $\log_{10} 2 = 0.30103\ldots$) Of course a better understanding of what is meant by the term *approximately* is required to make a deeper study of rational approximations of irrationals.

2.11: Let $s = \sqrt{2}+\sqrt{3}$. Then $s^2 = 5+2\sqrt{6}$ is irrational since $\sqrt{6}$ is irrational, and hence s itself must be irrational. But $s^4 = 49 + 20\sqrt{6}$. Thus $s^4 - 10s^2 + 1 = 0$. Hence s is algebraic.

3.1: Since $41 = 2^5 + 9$, the last person's position is $J(41) = 2 \cdot 9 + 1 = 19$.

3.2: The children from the first marriage were placed in positions 3, 7, 8, 9, 10, 11, 14, 15, 20, 22, 23, 24, 26, 27, and 30. The child in position 25 was chosen.

3.3: $J(50) = 37$, $J(199) = 143$, $J(512) = 1$, $J(1{,}000{,}000) = 951{,}425$.

3.4:

n	1	2	3	4	5	6	7	8	9	10	11	12	13	14	15	16	17	18	19	20	21	22	23	24	25
$J(n,4)$	1	1	2	2	1	5	2	6	1	5	9	1	5	9	13	1	5	9	13	17	21	3	7	11	15

3.5: $200 = 11001000_2$, $356 = 101100100_2$, $10{,}000 = 10011100010000_2$.

3.6: $1001001_2 = 73$, $10101110000_2 = 1{,}492$, $11011110000_2 = 1{,}776$.

3.7: $P(8,9) = \begin{pmatrix} 1 & 2 & 3 & 4 & 5 & 6 & 7 & 8 \\ 1 & 3 & 6 & 4 & 5 & 2 & 7 & 8 \end{pmatrix}.$

3.8: Shuffle n cards and let $X_i = 1$ if i^{th} card ends up in i^{th} position and $X_i = 0$ otherwise for $1 \leq i \leq n$. The expected value of X_i is $E(X_i) =$

$1(1/n) + 0(1 - 1/n) = 1/n$. Since expectation is linear, $E(\sum X_i) = \sum E(X_i) = n(1/n) = 1$.

3.9: $J(9,5) = 8$, $J(11,5) = 8$, $J(12,5) = 1$, $J(15,5) = 1$, $J(17,5) = 11$, $J(18,5) = 16$, $J(19,5) = 2$.

4.1: There are three alternative winning moves. Take one from the 8-pile, take three from the 9-pile, or take seven from the 11-pile.

4.2: Remove 28 counters from the 40-counter pile leaving 20 counters and 12 counters, respectively.

4.3: Every move will result in at least one 1 changing to a 0. In that column, the sum goes from even to odd.

4.4: For each pile there is at most one way to change all column sums to even numbers.

4.5: Make the total $100 - 11k$ for some k and then add $11 - r$ counters each time your opponent adds r counters.

4.6: Northcott's game is identical to Nim with eight piles, each with six counters.

4.7: **(a)** Leave $k(m+1)$ counters for some k. Then remove $m+1-r$ counters whenever your opponent removes r counters.

(b) In the misère version, leave $k(m+1)+1$ counters and proceed as in part (a).

4.8: Note that you can force multiples of seven unless a face value is depleted.

5.1: $2^{11} - 1 = 2{,}087 = 23 \cdot 89$. The converse of Observation 1 is not true.

5.2: $2^{23} - 1 = 47 \cdot 178{,}481$.

5.3: $a^n - 1 = (a-1)(a^{n-1} + a^{n-2} + \ldots + 1)$.

5.4: $\sigma(100) = 217, \sigma(1{,}000) = 2{,}340, \sigma(22{,}021) = \sigma(19^2 \cdot 61) = \sigma(19^2)$. $\sigma(61) = 381 \cdot 62 = 23{,}622$.

5.5: $2^{p-1}(2^p - 1) = \frac{n(n+1)}{2}$ for $n = 2^p - 1$.

5.6: Use induction to show that $1^3 + 3^3 + \ldots + (2n-1)^3 = 2^n(2^{n+1} - 1)$.

5.7: $\sigma(945) = 1{,}920 > 2 \cdot 945$.

5.8: $n = 2 : a = 220, b = 284$; $n = 4 : a = 17{,}296, b = 18{,}416$; $n = 7 : a = 9{,}363{,}584, b = 9{,}437{,}056$.

5.9: $s(2{,}620) = 2{,}924, s(6{,}232) = 6{,}368, s(122{,}368) = 123{,}152$.

5.10: (b) $12{,}496 \to 14{,}288 \to 15{,}472 \to 14{,}536 \to 14{,}264 \to 12{,}496$.

5.11: Use induction on n.

6.1: Let $S_n = a + ar + \ldots + ar^n = \frac{a(r^{n+1}-1)}{r-1}$. If $|r| < 1$, then $r^{n+1} \to 0$ as $n \to \infty$ and S_n converges to $\frac{a}{1-r}$.

6.2: Rewrite the sum so that each term has denominator equal to the least common multiple of 2, 3, ..., n. Let $2^m \leq n < 2^{m+1}$. Then each term will have an even numerator except the single term equal to $\frac{1}{2^m}$, which necessarily will have an odd numerator. So the sum of numerators will be odd and hence not divisible by the denominator. Hence the sum is not an integer.

6.3: The bug travels two miles since it takes one minute before the cars collide.

6.4: $\sum_{n=1}^{\infty} \frac{1}{(2n-1)^2} = \sum_{n=1}^{\infty} \frac{1}{n^2} - \sum_{n=1}^{\infty} \frac{1}{(2n)^2} = \frac{\pi^2}{8}$.

6.5: $1 = \sum_{n=1}^{\infty} (-1)^n \frac{\pi^{2n-1}}{2^{2n-1}(2n-1)!}$.

6.6: $B_1 = -\frac{1}{2}$, $B_2 = \frac{1}{6}$, $B_3 = 0$, $B_4 = -\frac{1}{30}$, $B_5 = 0$, $B_6 = \frac{1}{42}$, $B_7 = 0$, $B_8 = -\frac{1}{30}$.

6.7: $D_6 = 42$ since 6 is divisible by $2-1, 3-1$, and $7-1$.

6.8: $\zeta(2) = \frac{\pi^2}{6}$, $\zeta(4) = \frac{\pi^4}{90}$, $\zeta(6) = \frac{\pi^6}{945}$, $\zeta(8) = \frac{\pi^8}{9{,}450}$.

6.9: We get a "telescoping sum" where all but the first and last terms of the partial sums collapse.
$\sum_{n=1}^{\infty} \frac{1}{t_n} = \sum_{n=1}^{\infty} \frac{2}{n(n+1)} = 2 \cdot \lim_{N->\infty} \sum_{n=1}^{N} \frac{1}{n(n+1)}$
$= 2 \cdot \lim_{N->\infty} \sum_{n=1}^{N} [\frac{1}{n} - \frac{1}{n+1}] = 2 \cdot \lim_{N->\infty} (1 - \frac{1}{N+1}) = 2.$

7.1: Since $N \equiv 3 \pmod 4$, there must be a prime $p \equiv 3 \pmod 4$ that divides N. But p is not one of $p_1, \ldots, p_r$.

7.2: $3^{(17-1)/2} \equiv -1 \pmod{17}$.

7.3: 23

7.4: **(a)** $m_1 = 70, m_2 = 21, m_3 = 15$.

(b) 56

(c) $1{,}103 \equiv 53 \pmod{105}$.

7.5: 330

7.6: Southwest corner at (14, 20).

7.7: $\gcd(1274 + j, 1308 + k) > 1$ for $0 \le j \le 2, 0 \le k \le 2$.

8.1:
$$\begin{bmatrix} 30 & 39 & 48 & 1 & 10 & 19 & 28 \\ 38 & 47 & 7 & 9 & 18 & 27 & 29 \\ 46 & 6 & 8 & 17 & 26 & 35 & 37 \\ 5 & 14 & 16 & 25 & 34 & 36 & 45 \\ 13 & 15 & 24 & 33 & 42 & 44 & 4 \\ 21 & 23 & 32 & 41 & 43 & 3 & 12 \\ 22 & 31 & 40 & 49 & 2 & 11 & 20 \end{bmatrix}$$

8.2: There are no knight's tours on 3×3 or 4×4 chessboards, but there are on 5×5 chessboards as long as the knight starts in a corner.

8.3: The only Latin squares of order 2 are $\begin{bmatrix} 1 & 2 \\ 2 & 1 \end{bmatrix}$ and $\begin{bmatrix} 2 & 1 \\ 1 & 2 \end{bmatrix}$. They are not orthogonal.

8.4: Start with the orthogonal pair $A = \begin{bmatrix} 1 & 2 & 3 \\ 2 & 3 & 1 \\ 3 & 1 & 2 \end{bmatrix}$ and

$B = \begin{bmatrix} 1 & 2 & 3 \\ 3 & 1 & 2 \\ 2 & 3 & 1 \end{bmatrix}$ and transpose the same pair of rows or columns in both A and B.

8.5: (b) $A_1 = \begin{bmatrix} 1 & 2 & 3 & 4 & 5 \\ 2 & 3 & 4 & 5 & 1 \\ 3 & 4 & 5 & 1 & 2 \\ 4 & 5 & 1 & 2 & 3 \\ 5 & 1 & 2 & 3 & 4 \end{bmatrix}$, $A_2 = \begin{bmatrix} 1 & 2 & 3 & 4 & 5 \\ 3 & 4 & 5 & 1 & 2 \\ 5 & 1 & 2 & 3 & 4 \\ 2 & 3 & 4 & 5 & 1 \\ 4 & 5 & 1 & 2 & 3 \end{bmatrix}$,

$A_3 = \begin{bmatrix} 1 & 2 & 3 & 4 & 5 \\ 4 & 5 & 1 & 2 & 3 \\ 2 & 3 & 4 & 5 & 1 \\ 5 & 1 & 2 & 3 & 4 \\ 3 & 4 & 5 & 1 & 2 \end{bmatrix}$, $A_4 = \begin{bmatrix} 1 & 2 & 3 & 4 & 5 \\ 5 & 1 & 2 & 3 & 4 \\ 4 & 5 & 1 & 2 & 3 \\ 3 & 4 & 5 & 1 & 2 \\ 2 & 3 & 4 & 5 & 1 \end{bmatrix}$

8.6: Let the columns represent positions and the numbers the tires.

8.7:

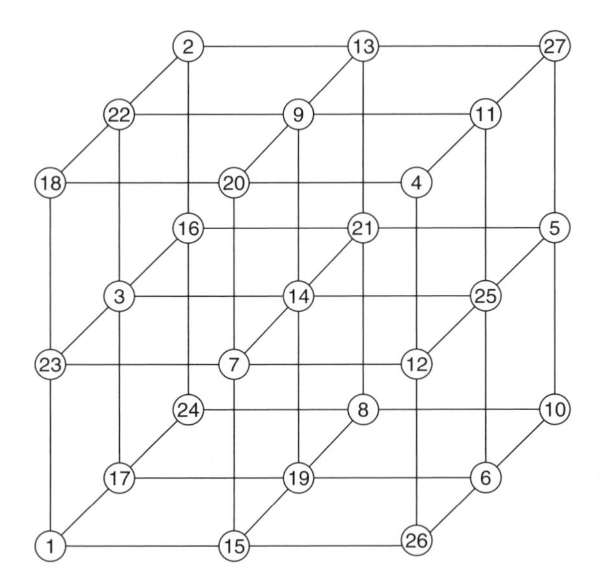

Figure A.5 A $3 \times 3 \times 3$ magic cube.

8.8: Here is one option for 16 golfers 1–16:

Day 1—(1 2 3 4), (5 6 7 8), (9 10 11 12), (13 1 4 15 16), Day 2—(1 5 9 13), (2 6 10 14), (3 7 11 15), (4 8 12 16),

Day 3—(1 6 11 16), (2 5 12 15), (3 8 9 14), (4 7 10 13), Day 4—(1 7 12 14), (2 8 11 13), (3 5 10 16), (4 6 9 15),

Day 5—(1 8 10 15), (2 7 9 16), (3 6 12 13), (4 5 11 14).

8.9: $D_2 = 1$, $D_3 = 2$, $D_4 = 9$, $D_5 = 44$.

8.10: **(a)** For every *normalized* Latin square with first row and column in consecutive numerical order, every rearrangement of the n columns followed by rearrangement of all but the first row results in a distinct Latin square. There are $n!(n-1)!$ ways to do this.

(b) $l_1 = 1, l_2 = 1, l_3 = 1, l_4 = 4$.

9.1: $p_1 = P_1 = \frac{1}{4}$, $p_2 = P_2 = \frac{1}{8}$, $p_3 = P_3 = \frac{3}{32}$, $p_4 = P_4 = \frac{5}{64}$, $p_5 = P_5 = \frac{35}{512}$, $p_6 = P_6 = \frac{217}{512}$.

9.2: Choose the red die since its expectation for each roll is $\frac{10}{3}$, the highest of the four dice.

9.3: In this case, with proper choice of dice, your chance of winning goes up to $\frac{20}{27}$. Verify by calculating the probability of WW, WLW, or LWW where $p(W) = \frac{2}{3}$ and $p(\text{L}) = \frac{1}{3}$.

9.4: 24

9.5: Either (1, 2, 4, 5) and (1, 2, 3, 3, 4, 5, 5, 6, 7) or (1, 4, 4, 7) and (1, 2, 2, 3, 3, 3, 4, 4, 5) have standard probability distribution.

9.6: In this case, not all generating functions would require a factor of x.

9.7: These two weighted dice have a fair sum.

9.8: $p(R < r) = \frac{1}{3}$, $p(R = r) = \frac{1}{6}$.

9.9: **(a)** $\frac{1}{2}$

(b) $\frac{7}{12}$

9.10: One possibility is the following three dice:

	2	
6	2	6
	2	
	2	

,

	4	
4	4	4
	4	
	4	

,

	5	
1	5	1
	5	
	5	

.

10.1: $2n$

10.2: No, for all integers n, color points in the intervals $[2n, 2n + 1)$ blue and color all points in the intervals $[2n - 1, 2n)$ red.

10.3: **(a)** Yes. Consider the vertices of a unit tetrahedron.

(b) Yes. Mimic solution to Problem #3 using a unit tetrahedron.

10.4: In Figure A.6, three lines through the chosen point parallel to the three sides of the triangle have been added. But now $\text{A} + \text{C} + \text{E} = (\text{r} + \text{s}) + (\text{t} + \text{u}) + (\text{v} + \text{w}) = (\text{s} + \text{t}) + (\text{u} + \text{v}) + (\text{w} + \text{r}) = \text{B} + \text{D} + \text{F}$.

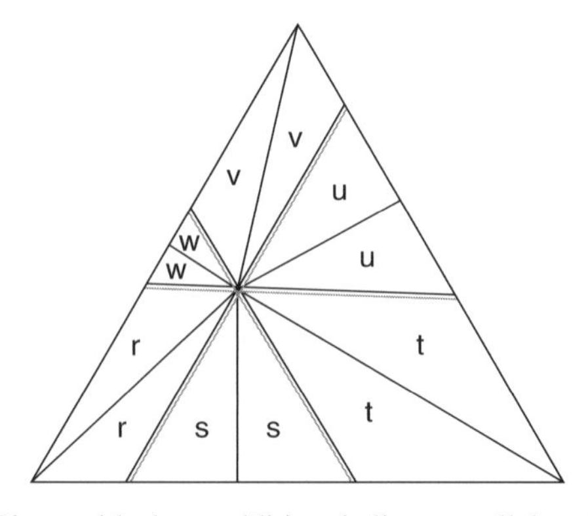

Figure A.6 Pizza with three additional slices parallel to the three sides.

10.5: Nine line segments can be colored black without forming a black triangle as shown in the figure below.

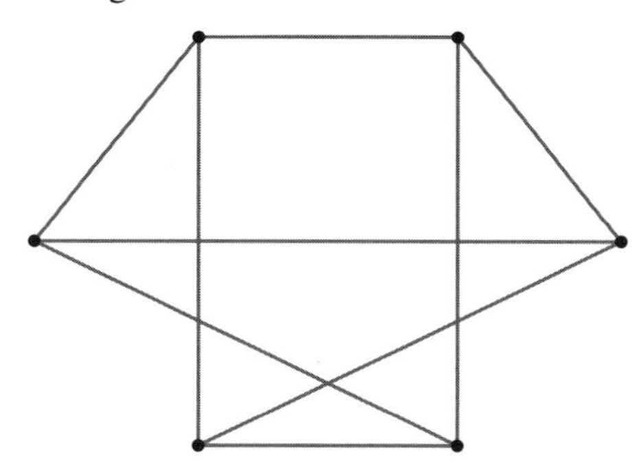

Figure A.7 Nine line segments with no black triangle.

10.6: Yes, cut the bar one-seventh from the left and two-sevenths from the right. Then every multiple of one-seventh from one-seventh up to seven-sevenths can be dispensed.

11.1: $3\frac{1}{8} = 3.125$, $\frac{256}{81} = 3.16049382716\ldots$, $\sqrt{10} = 3.16227766017\ldots$, $4 \cdot (\frac{9785}{11136})^2 = 3.08832649\ldots$, $\frac{377}{120} = 3.141666666\ldots$, $\frac{22}{7} = 3.142857142857\ldots$, $\frac{355}{113} = 3.14159292035\ldots$.

11.2: For a unit circle, the inscribed square has sides of length $\sqrt{2}$ and the circumscribed hexagon has sides of length $\frac{2\sqrt{3}}{3}$. So the square has perimeter $4\sqrt{2}$, while the hexagon has perimeter $4\sqrt{3}$. Now take the mean to approximate the circumference of the circle.

11.3: **(a)** $10\pi \approx 4[2\sqrt{10} + \sqrt{2}]$. So $\pi \approx 3.09550755\ldots$.

(b) $50\pi \approx 4[15\sqrt{2} + 8\sqrt{5}]$. So $\pi \approx 3.12814\ldots$.

(c) $250\pi \approx 4[125\sqrt{2} + 6\sqrt{10}]$. So $\pi \approx 3.13200\ldots$.

11.4: **(a)** Let $A = B = \frac{x}{2}$ in the formula $\sin(A + B) = \sin A \cos B + \cos A \sin B$.

(b) Let $A = B = \frac{x}{2}$ in the formula $\cos(A + B) = \cos A \cos B - \sin A \sin B$. Then the use the fact that $\cos^2(\frac{x}{2}) = 1 - \sin^2(\frac{x}{2})$.

11.5: The length of the rectangle is the circumference of the wheel, namely $2\pi r$.

11.6: **(a)** Take the tangent of both sides of the arctangent formula and use the fact that tangent and arctangent are inverse functions on appropriately restricted domains.

(b) $\frac{\pi}{4} = \arctan(1) = \arctan(\frac{1}{2}) + \arctan(\frac{1}{3})$

(c) $\arctan(\frac{1}{2}) = \arctan(\frac{1}{3}) + \arctan(\frac{1}{7})$. Now substitute this into the answer in part (b).

11.7: $a_1 = 1.20710678118$, $b_1 = 1.189207115$, $a_2 = 1.19815694809$, $b_2 = 1.19812352149$, $a_3 = 1.19814023479$, $b_3 = 1.19814023467$.

11.8: **(a)** 15 decimal place accuracy

(b) 8 decimal place accuracy

11.9: $\sqrt[6]{\pi^4 + \pi^5}$ is approximately e (the base of the natural logarithm function) to 8 decimal place accuracy. Coincidence? I *do* think so!

11.10: The error caused by truncating a convergent alternating series is less than the first neglected term. Let $p_k = \pi + \epsilon$ where ϵ is the error in using p_k to estimate π. Then $p_{k+1} = p_k + \sin p_k = (\pi + \epsilon) + \sin(\pi + \epsilon) = \pi + \epsilon - \sin\epsilon = \pi + \epsilon - (\frac{\epsilon}{1!} - \frac{\epsilon^3}{3!} + \frac{\epsilon^5}{5!} - \ldots) = \pi + \frac{\epsilon^3}{6} - \ldots$ Hence if $\epsilon < 10^{-n}$, then $\frac{\epsilon^3}{6} < 0.2 \cdot 10^{-3n}$. It follows that p_{k+1} has at least three times as many accurate decimal places as does p_k.

11.11: $c_1 = \frac{22}{7}$, $c_2 = \frac{333}{106}$, $c_3 = \frac{355}{113}$, $c_4 = \frac{103993}{33102}$, $c_5 = \frac{104348}{33215}$, $c_6 = \frac{208341}{66317}$, $c_7 = \frac{312689}{99532}$, $c_8 = \frac{833719}{265381}$, $c_9 = \frac{1146408}{364913}$, $c_{10} = \frac{4272943}{1360120}$. The tenth convergent provides 13 decimal place accuracy to π.

12.1: The same proof works for any point $p = (\sqrt{n}, \frac{1}{r})$ for any nonsquare n and $r \geq 3$ as well as for many other carefully chosen points.

12.2: For n sets, let A_i consist of the entries in all rows congruent to i modulo n for $1 \leq i \leq n$.

12.3: **(a)** Since all squares are congruent to either 0, 1, or 4 (mod 8), the sum of three such squares could never be congruent to 7 (mod 8). Hence all integers that are congruent to 7 (mod 8) cannot equal the sum of three nonzero squares.

(b) Notice that there is a one-to-one correspondence between representations of $2n$ as a sum of four squares (not all nonzero necessarily) and representations of $8n$ as a sum of four squares. In one direction, if $2n = a^2 + b^2 + c^2 + d^2$, then $8n = (2a)^2 + (2b)^2 + (2c)^2 + (2d)^2$. In the other direction, if $8n + A^2 + B^2 + C^2 + D^2$, then each of A, B, C, D must be even (since odd squares are congruent to 1 modulo 8). But then $2n = (A/2)^2 + (B/2)^2 + (C/2)^2 + (D/2)^2$, a sum of integral squares. Since 2 is not representable as the sum of four nonzero squares, neither are the numbers $2 \cdot 4^r$ for all $r \geq 1$.

12.4: The solution to Problem #3 provides a proof for $k = 5$ where it is shown that any integer greater than 33 can be expressed as the sum of five nonzero squares. If $k \geq 6$ and $n \geq 29 + k$, then $n + 5 - k \geq 34$. Thus $n + 5 - k = x_1^2 + x_2^2 + x_3^2 + x_4^2 + x_5^2$ for nonzero x_i. But then $n = x_1^2 + x_2^2 + x_3^2 + x_4^2 + x_5^2 + 1^2 + \ldots + 1^2$ with $k - 5$ terms of 1^2. The result is a representation of n as the sum of k nonzero squares.

12.5: Verify that if $34 \leq n \leq 78$, then n is representable as the sum of distinct triangular numbers less than or equal to 36. Now add 45 to the representations for 34, 35, ..., 78 to obtain expressions for n with $79 \leq n \leq 123$ as the sum of distinct triangular numbers. Next add 55 to the representations for 69, 70, ..., 123 to get expressions for n with $124 \leq n \leq 178$ as the sum of distinct triangular numbers. Next add 66 to the representations for

113, 114, ..., 178, and so on. This algorithm extends indefinitely since each triangular number ≥ 6 is less than twice the one before.

12.6: Let the set consist of the numbers $a_1, \ldots, a_{2,005}$. Consider the particular sums $s_1 = a_1, s_2 = a_1 + a_2, \ldots, s_{2,005} = a_1 + \ldots + a_{2,005}$ and form the set $\{0, s_1, s_2, \ldots, s_{2,005}\}$. By the pigeonhole principle, two elements of the set belong to the same congruence class modulo 2,005. If $s_i \equiv s_j \pmod{2{,}005}$ with $i < j$, then 2,005 divides $s_j - s_i = a_{i+1} + \ldots + a_j$. If $0 \equiv s_i \pmod{2{,}005}$ for some i, then s_i is the appropriate sum.

12.7: Mentally slice the cube with three orthogonal slices down the center of each side, thereby creating eight unit subcubes. By the pigeonhole principle, two of the points lie in or on the same subcube. But the farthest apart they could be would occur if they were at opposite diagonal corners. By applying the Pythagorean Theorem twice, it is readily seen that the opposing corners are $\sqrt{3}$ units apart.

12.8: The number $12{,}167 = 23^3$ and $12{,}168 = 2^3 \cdot 3^2 \cdot 23^2$.

13.1: $\phi(3) = 2, \phi(5) = 4, \phi(15) = 8, \phi(24) = 8, \phi(48) = 16, \phi(101) = 100, \phi(105) = 48, \phi(1{,}000) = 400$.

13.2: **(a)** Since m and n are relatively prime, if $r|mn$, then there are unique integers $d|m$ and $e|n$ for which $r = de$.

(b) Note that $\phi(p) = p - 1$ since every integer less than a prime p is relatively prime to it. Similarly, every number that is not a multiple of p is relatively prime to p^t. Hence $\phi(p^t) = p^t - \frac{1}{p} \cdot p^t = p^{t-1}(p-1)$.

(c) The number of integers less than or equal to one billion that are not relatively prime to one billion is $10^9 - \phi(10^9) = 10^9 - \phi(2^9 5^9) = 10^9 - 2^8 5^8 \cdot 4 = (10 - 4)10^8 = 600{,}000{,}000$.

13.3: By the Euler-Fermat Theorem, if $\gcd(10, n) = 1$, then $10^{\phi(n)} \equiv 1 \pmod{n}$. Hence n divides the number consisting of a string of $\phi(n)$ nines. Of course n will also divide a string of $k \cdot \phi(n)$ nines for any $k \geq 1$.

13.4: Consider the n fractions $\frac{1}{n}, \ldots, \frac{n}{n}$. Reduce each one to lowest terms. For each $d|n$, there are $\phi(d)$ reduced fractions with denominator d.

13.5: **(a)** $15{,}841 = 7 \cdot 31 \cdot 73$

(b) $41{,}041 = 7 \cdot 11 \cdot 13 \cdot 41$

13.6: **(a)** $2^{66} \equiv 1 \pmod{67}$, $2^{33} \equiv -1 \pmod{67}$, $2^{22} \equiv 37 \pmod{67}$, and $2^6 \equiv -3 \pmod{67}$. By Lucas's Primality Test, 67 is prime.

(b) $2^{90} \equiv 64 \not\equiv 1 \pmod{91}$. Hence 91 cannot be prime by Fermat's Little Theorem. If we had chosen base $b = 3$ instead though, then $3^{90} \equiv 1 \pmod{91}$. Furthermore, $3^{45} \not\equiv 1 \pmod{91}$. However, $3^{30} \equiv 1 \pmod{91}$. Hence Lucas's Primality Test would confirm that 91 is composite.

13.7: $1{,}105 = 5 \cdot 13 \cdot 17$. Hence reason modulo 5, 13, and 17. $2^{1,104} = (2^4)^{276} \equiv 1^{276} = 1 \pmod{5}$, $2^{1,104} = (2^6)^{184} \equiv (-1)^{184} \equiv 1 \pmod{13}$, and $2^{1,104} = (2^4)^{276} \equiv (-1)^{276} = 1 \pmod{17}$. Thus $2^{1,104} \equiv 1 \pmod{1105}$

since 5, 13, and 17 are pairwise relatively prime. Hence 1105 is a psp(2). In addition, $3^{1,104} = (3^4)^{276} \equiv 1^{276} = 1 \pmod 5$, $3^{1,104} = (3^3)^{368} \equiv 1^{368} = 1 \pmod{13}$, and $3^{1,104} = (3^8)^{138} \equiv (-1)^{138} = 1 \pmod{17}$. Hence $3^{1,104} \equiv 1 \pmod{1,105}$ and so 1,105 is a psp(3).

13.8: **(a)** Since r is composite, there exist integers a, b with $1 < a \leq b < r$ such that $r = ab$. Hence $r' = 2^r - 1 = 2^{ab} - 1 = (2^a - 1)(2^{a(b-1)} + \ldots + 2^a + 1)$. Thus r' is composite.

(b) Since $2^{r-1} \equiv 1 \pmod r$, $2^{r-1} - 1 = rd$ for some integer d. Hence $2^{r'-1} = 2^{2^r-2} = 2^{2(2^{r-1}-1)} = 2^{2rd} = (2^r)^{2d} \equiv 1^{2d} = 1(\text{mod} r')$.

(c) Since 341 is a pseudoprime, the set of pseudoprimes is nonempty. If r is a pseudoprime, then $r' = 2^r - 1$ is a pseudoprime. Hence we can construct pseudoprimes indefinitely.

14.1: The number $N = 4p_1 \cdots p_n - 1 \equiv 3 \pmod 4$. But the product of integers all of which are congruent to 1 modulo 4 is itself congruent to 1 modulo 4. So N must be divisible by a prime $p \equiv 3 \pmod 4$. But p cannot be among $p_1, \ldots, p_n$ since none of them are divisors of N.

14.2: Reason why the list of primes 2, 3, 5, 7, 13, 23, 43, 83, 163, 317, 631, and 1,259 establishes Bertrand's Postulate for $1 < n < 1{,}000$.

14.3: **(a)** Let $\sigma(n)$ be the sum of the divisors of n. Then $\sigma(945) = 1{,}920 > 2 \cdot 945$.

(b) Note that if d is a divisor of n, then $3d$ is a divisor of $3n$. Hence $\sigma(3n) \geq 3\sigma(n)$. Since 945 is abundant, the numbers $3^n \cdot 945$ must necessarily be abundant as well.

14.4: $\sigma(6) = \sigma(11) = 12$.

14.5: **(a)** Given an integer n and $k > 1$, note that if d is a divisor of n, then kd is a divisor of kn. In addition,1 is a divisor of kn not of the form kd. Hence $\sigma(kd) > k\sigma(n)$. If $\sigma(n) \geq 2n$, then $\sigma(kn) > 2kn$.

(b) The abundant numbers less than 46 are 12, 18, 20, 24, 30, 36, 40, and 42. No two sum to 46.

(c) Let n be an even number greater than 46. If $n \equiv 0 \pmod 6$, then $n = 12+(n-12)$ is such a representation. If $n \equiv 2 \pmod 6$, then write $n = 20 + (n - 20)$. If $n \equiv 4 \pmod 6$, then write $n = 40 + (n - 40)$.

14.6: For example, for $p = 89 : 86 = 89{-}3, 84 = 89{-}5, 82 = 89{-}7$, and even integers less than or equal to 80 have already been verified for $p = 83$.

14.7: **(a)** $\frac{4}{2} = \frac{1}{1} + \frac{1}{2} + \frac{1}{2}$, $\frac{4}{3} = \frac{1}{1} + \frac{1}{4} + \frac{1}{12}$, $\frac{4}{4} = \frac{1}{2} + \frac{1}{3} + \frac{1}{6}$, $\frac{4}{5} = \frac{1}{2} + \frac{1}{4} + \frac{1}{20}$, $\frac{4}{6} = \frac{1}{2} + \frac{1}{9} + \frac{1}{18}$, $\frac{4}{7} = \frac{1}{2} + \frac{1}{21} + \frac{1}{42}$, $\frac{4}{8} = \frac{1}{3} + \frac{1}{9} + \frac{1}{18}$, $\frac{4}{9} = \frac{1}{3} + \frac{1}{12} + \frac{1}{24}$, $\frac{4}{10} = \frac{1}{3} + \frac{1}{45} + \frac{1}{90}$, $\frac{4}{11} = \frac{1}{3} + \frac{1}{99} + \frac{1}{198}$, $\frac{4}{12} = \frac{1}{9} + \frac{1}{9} + \frac{1}{9}$.

(b) $\frac{5}{2} = \frac{1}{1} + \frac{1}{1} + \frac{1}{2}$, etc.

14.8: 3, 1, 5, 4, 6, 2.

14.9: 70 is a weird number since it is abundant, but no subset of its proper divisors sum to 70. (Notice that the word "weird" is weird as well in that it does not obey the usual "i before e" rule.)

14.10: **(a)** If the numbers $n+1, \ldots, 2n$ are chosen then none divides any other. Hence the answer must be at least n. Interestingly, Erdös showed this is the actual answer. Let T be the largest subset of S with no member dividing another. Consider the $[\frac{n+1}{2}]$ pair of numbers $(2n, n)$, $(2n-2, n-1), \ldots (2[\frac{n+2}{2}], [\frac{n+2}{2}])$. The set T can contain at most one member from each pair. To maximize its size, assume that it does. In addition, T cannot contain any of the numbers 1, 2, $\ldots, [\frac{n}{2}]$ since each of them divides evenly into some number of the set $\{[\frac{n+2}{2}], \ldots, n\}$. But for each such number, either T contains a multiple of it among $\{[\frac{n+2}{2}], \ldots, n\}$ or double that number. Finally, T can contain all $[\frac{n}{2}]$ odd numbers $n+1, n+3, \ldots, 2n+1$. Thus T contains at most $[\frac{n+1}{2}] + [\frac{n}{2}] = n$ elements.

(b) No one knows the answer. By taking the numbers $m+1, m+2, \ldots, 3m+2$ for m as large as possible, we see that the answer must be at least $[2n/3]$. In some instances, the answer seems a bit larger. For example, for $n = 29$, D.J. Kleitman has shown that the set 11, 12, ..., 29 with 18 and 24 omitted and then 6, 8, 9, and 10 appended satisfies the conditions. There's certainly more to learn.

15.1: Use Ferrers's graphs and consider the conjugate partition.

15.2: There is a one-to-one correspondence between partitions of n containing at least one part k and partitions of $p(n-k)$.

15.3: **(a)** $\prod_{k=1}^{3} \frac{1}{1-x^k} = \sum_{n=0}^{\infty} p(n, 3)x^n$.

(b) $p(n, 3) = 1+x+2x^2+3x^3+4x^4+5x^5+7x^6+\ldots$. Hence $p(6, 3) = 7$. The partitions are $3+3$, $3+2+1$, $3+1+1+1$, $2+2+2$, $2+2+1+1$, and $2+1+1+1+1$, and $1+1+1+1+1+1$

15.4: **(a)**

n	1	2	3	4	5	6	7	8	9	10
$p(n, e)$	0	1	0	2	0	3	0	5	0	7

(b) $\prod_{k=1}^{\infty} \frac{1}{1-x^{2k}} = \sum_{n=0}^{\infty} p(n, e)x^n$.

(c) $p(n, e) = p(n/2)$ for all even n.

15.5: $\prod_{k=1}^{\infty}(1 + x^{2k}) = \sum_{n=0}^{\infty} p_u(n, e)x^n, \prod_{k=1}^{\infty}(1 + x^{2k-1}) = \sum_{n=0}^{\infty} p_u(n, o)x^n$.

15.6: **(a)**

n	1	2	3	4	5	6	7	8	9	10
$p_u(n, o)$	1	0	1	1	1	1	1	2	2	2

(b) $p_u(n, o) = sc(n)$ for all n. To see why, do the following. For any partition p of n into unequal odd parts, create a new Ferrers's graph of n by laying out the parts of p in nested L-shaped strings. Figure A.8 shows the transformation for the three relevant partitions of $n = 12$.

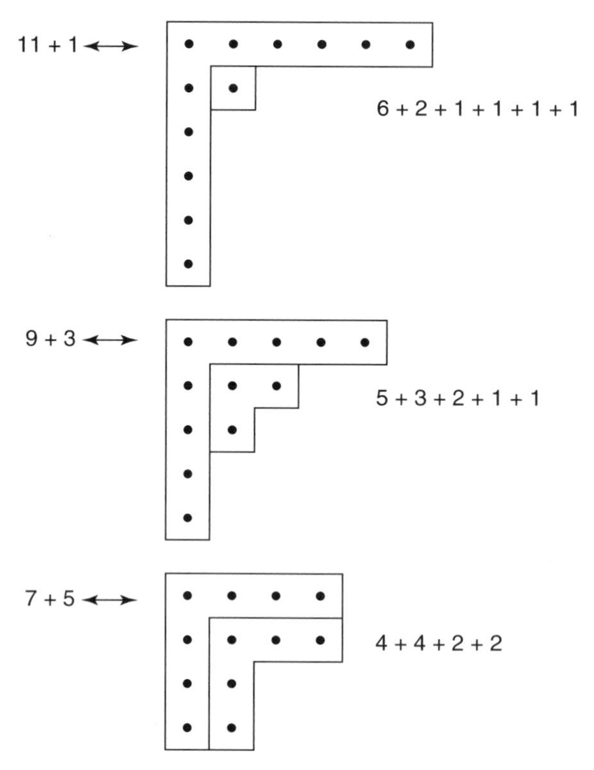

Figure A.8 $p_u(12, \sigma) = sc(12)$.

15.7: Let $C(n)$ = the number of compositions of n with no part 1. Note that $C(2) = 1 = F_1$. It suffices to show that $C(n) = C(n-1) + C(n-2)$ for $n \geq 3$ since that is the same linear recurrence satisfied by the Fibonacci numbers. For arbitrary n, let $r_1, \ldots, r_i$ be the compositions of $n-2$ with no part 1, and let $s_1, \ldots, s_j$ be the compositions of $n-1$ with no part 1. By adding "+2" to the right of each of the r's, we get compositions of n with no part 1. (For example, $3+4+5$ becomes $3+4+5+2$.) Also by adding one to the right-most part of each of the s's gives other compositions of n with no part 1. (For example, $4+3+6$ becomes $4+3+7$.) The two sets of compositions of n are disjoint since the first set has right-most part 2, while none of the second set do. Furthermore, these are all the compositions of n with no part 1. Hence $C(n) = C(n-1) + C(n-2)$ as desired.

15.8: **(a)** $p(9) = p(8) + p(7) - p(4) - p(2) = 22 + 15 - 5 - 2 = 30$.

(b) $p(10) = p(9) + p(8) - p(5) - p(3) = 30 + 22 - 7 - 3 = 42$.

15.9: n-uplets come in $p(n)$ different varieties.

16.1: For each k with $0 \leq k \leq n$, there are $\binom{n}{k}$ ways to choose k cards for Player #1. But each hand choice for Player #1 corresponds to a unique hand for Player #2 made up of the remaining cards of the deck. Hence the total number of ways to deal two hands is $\sum_{k=0}^{n} \binom{n}{k} = 2^n$. If each player must receive at least one card, then the answer would be $2^n - 2$.

16.2: There are m^n ways to distribute n cards among m players. This follows from a multinomial theorem, a natural extension of the binomial theorem.

16.3: Yes, suppose that $\binom{n}{r}, \binom{n}{r+1}, \binom{n}{r+2}$ are three consecutive entries in the ratio 1:2:3. Then $\binom{n}{r+2} = 3\binom{n}{r}$ and so $r!(n-r)! = 3(n-r-2)!(r+2)!$ after clearing denominators. Thus $(n-r)(n-r-1) = 3(r+1)(r+2)$. But $\binom{n}{r+1} = 2\binom{n}{r}$, and so $n - r = 2(r+1)$. That is, $n = 3r + 2$. Thus $(2r+2)(2r+1) = 3(r+1)(r+2)$ and we get that $r = 4$. It follows that $n = 14$. Hence the consecutive binomial coefficients $\binom{14}{4} = 1{,}001$, $\binom{14}{5} = 2{,}002$, and $\binom{14}{6} = 3{,}003$ are in the ratio 1:2:3.

16.4: The number of ways of placing the 20 balls into the six numbered boxes is the same as the number of sequences of 20 1's, 2's, ..., 6's containing all six numbers and written in nondecreasing order. But the latter can be described by placing 20 markers and then placing 5 dividers among the 19 spaces between the markers. There are $\binom{19}{5}$ such choices.

16.5: This solution to this problem is analogous to that of Problem 16.4. Think of the candy bars as being the balls and the children as being the boxes. The answer is $\binom{39}{9}$.

16.6: For $1 \leq k \leq n-1$, $\binom{n}{k} = \frac{n(n-1)\cdots(n-k+1)}{k!}$. If n is prime, then n divides the numerator but not the denominator. Hence $n | \binom{n}{k}$. If n is composite, then we can write $n = rs$ where $1 < r \leq s < p$. Then $\binom{n}{r} = \frac{rs(rs-1)\cdots(rs-r+1)}{r!} = \frac{s(rs-1)\cdots(rs-r+1)}{(r-1)!}$. But r does not divide any of the factors $rs-r+1, \ldots, rs-1$. Hence rs cannot divide the numerator. Therefore $\binom{n}{r}$ is not divisible by n.

16.7: Each south-east spelling of MATHEMATICS corresponds to a ten-letter word consisting of five letters S and five letters E corresponding to the five times we move south and five times we move east. The number of such words is $\frac{10!}{5!5!} = 252$.

16.8: $\sum_{k=1}^{n}(2k-1)^2 = \sum_{k=1}^{2n} k^2 - \sum_{k=1}^{n}(2k)^2 = \sum_{k=1}^{2n} k^2 - 4\sum_{k=1}^{n} k^2 = \frac{2n(2n+1)(4n+1)}{6} - \frac{4n(n+1)(2n+1)}{6} = \frac{2n(2n+1)[4n+1-2(n+1)]}{6} = \frac{(2n+1)(2n)(2n-1)}{6} = \binom{2n+1}{3}$.

16.9: Many combinatorial identities can be best understood by telling an appropriate story. Suppose you have $n+m$ distinguishable balls and wish to choose r of them. The number of ways to do so is $\binom{n+m}{r}$. But one method to make a selection is to first put n of the balls in one container and m in another. Then for some k with $0 \le k \le r$, choose k balls from the first container. Having done that, choose the rest ($r-k$ balls necessarily) from the second container. The total number of such selections is $\sum_{k=0}^{r}\binom{n}{k}\binom{m}{r-k}$.

16.10: We establish the result by induction on n. For $n = 1$, we have that $\sum_{k=0}^{1}\binom{1}{k} = \binom{2}{1}$. Now assume that the proposition is true for arbitrary n. We will show that $\sum_{k=0}^{n+1}\binom{n+1}{k}^2 = \binom{2n+2}{n+1.}$. Since $\binom{n+1}{k} = \binom{n}{k-1} + \binom{n}{k}$, it follows that $\sum_{k=0}^{n+1}\binom{n+1}{k}^2 = \sum_{k=0}^{n+1}\left(\binom{n}{k-1}+\binom{n}{k}\right)^2$ where we define $\binom{n}{-1} = \binom{n}{n+1} = 0$. Thus $\sum_{k=0}^{n+1}\binom{n+1}{k}^2 = \sum_{k=0}^{n+1}\binom{n}{k-1}^2 + \sum_{k=0}^{n+1}\binom{n}{k}^2 + \sum_{k=0}^{n+1} 2\binom{n}{k-1}\binom{n}{k} = 2\binom{2n}{n} + \sum_{k=0}^{n+1} 2\binom{n}{k-1}\binom{n}{k}$ by our inductive hypothesis.

But $\binom{n}{k-1} = \binom{n}{n-k+1}$ for all k with $0 \le k \le n+1$. Thus $\sum_{k=0}^{n+1} 2\binom{n}{k-1}\binom{n}{k} = \sum_{k=0}^{n+1} 2\binom{n}{k}\binom{n}{n-k+1}$. We now apply the result from Problem 16.9 with $m = n$ and $r = n+1$ to obtain $\sum_{k=0}^{n+1} 2\binom{n}{k}\binom{n}{n-k+1} = 2\binom{2n}{n+1}$. Therefore, $\sum_{k=0}^{n+1}\binom{n+1}{k}^2 = 2\binom{2n}{n} + 2\binom{2n}{n+1} = 2\binom{2n+1}{n+1} = \frac{2(2n+1)!}{n!(n+1)!} = \frac{2(n+1)(2n+1)!}{(n+1)!(n+1)!} = \binom{2n+2}{n+1}$.

Bibliography

Aigner, M., Ziegler, G.M. *Proofs from the Book*. Berlin: Springer, 1999.

Alford, W.R., Granville, A., and Pomerance, C. "There are Infinitely Many Carmichael Numbers." *Annals of Mathematics* 140, 1994: 703–722.

Arndt, J. and Haenel, C. (transl. Lishka C. and D.) *Pi—Unleashed*. Berlin: Springer, 2001.

Ayoub, R. "Euler and the Zeta Function." *American Mathematical Monthly* December, 1974: 1067–1086.

Bailey, D.H., Borwein, J.M., Borwein, P.B. and Plouffe, S. "The Quest for Pi", *The Mathematical Intelligencer*, Vol. 19, No. 1, 1997: 50–56.

Baillie, R., "Sums of Reciprocals of Integers Missing a Given Digit", *American Mathematical Monthly*, Vol. 86, 1979, 372–374.

Beckman, P. *A History of π*. New York: St. Martin's, 1971.

Benjamin, A.T. and Quinn, J.J. *Proofs that Really Count: The Art of Combinatorial Proof*. Washington D.C.: Mathematical Association of America, 2003.

Behforooz, G.H. "Thinning Out the Harmonic Series." *Math. Magazine*, Vol. 68, 1995, 289–293.

Boyer, C. and Merzbach, U.C. *A History of Mathematics* 2d ed. New York: Wiley, 1989.

Bressoud, D.M., *Factorization and Primality Testing*. New York: Springer, 1989.

Caldwell, C. "The Prime Pages", http://primes.utm.edu/research/primes, last accessed 8/4/03.

Castellanos, D. "The Ubiquitous π (Part I)." *Math. Magazine* Vol. 61, No. 2, April, 1998: 67–100.

Castellanos, D. "The Ubiquitous π (Part II)." *Math. Magazine* Vol. 61, No. 3, June, 1998: 148–163.

Chung, F. and Graham, R. *Erdös on Graphs—His Legacy of Unsolved Problems*. Wellesley, MA: AK Peters, 1998.

Cipra, B. *What's Happening in the Mathematical Sciences (Volume 1)* American Mathematical Society, 1993.

Dickson, L.E. *History of the Theory of Numbers* 3 vols. New York: Chelsea, 1966.

Gardner, M. *Mathematical Carnival*. Washington D.C.: Mathematical Association of America, 1989.

Guy, R.K. ed. *Reviews in Number Theory 1973–1983*. 6 vols. Providence: American Mathematical Society, 1984.

Guy, R.K. *Unsolved Problems in Number Theory*. 2d ed. New York: Springer, 1994.

Hardy, G.H. *Ramanujan*. 3d ed., New York: Chelsea, 1978.

Honsberger, R. (ed.) *Mathematical Plums*. MAA, 1979.

Mathematical Journeys, by Peter D. Schumer
ISBN 0-471-22066-3

Honsberger, R. *From Erdös to Kiev—Problems of Olympiad Caliber*. Mathematical Association of America, 1996.

Katz, V. *A History of Mathematics—An Introduction* 2d ed. Reading, MA: Addison Wesley, 1998.

Konhauser, J.D.E., Velleman, D., and Wagon, S. *Which Way Did the Bicycle Go?* Washington D.C.: Mathematical Association of America, 1996.

Kürschák, J. (transl. Rapaport. E.) *Hungarian Problem Book I*. New York: Random House, 1963.

LeVeque, W.J. ed. *Reviews in Number Theory*. 6 vols. Providence: American Mathematical Society, 1974.

Mann, H. and Shanks, D. "A Necessary and Sufficient Condition for Primality, and its Source." *J. Combinatorial Theory, Series A*, 13, July, 1972: 131.

Marcus, D.A. *Combinatorics: A Problem Oriented Approach*. Washington D.C.: Mathematical Association of America, 1998.

Monastrysky, M. *Modern Mathematics in the Light of the Fields Medals*. Wellesley, MA: A.K. Peters, 1998.

Pomerance, C., Selfridge, J.L, and Wagstaff, S.S. "The Pseudoprimes to $25 \cdot 10^9$." *Mathematics of Computation* 35, no. 151, July, 1980: 1003–1026.

Ribenboim, P. *Catalan's Conjecture*. Boston: Academic Press, 1994.

Ribenboim, P. *The New Book of Prime Number Records*. New York: Springer, 1996.

Sárközy, A. "On Divisors of Binomial Coefficients, I." *Journal of Number Theory* 20, no. 1, February, 1985: 70–80.

Schumer, P.D. *Introduction to Number Theory*. Boston: PWS, 1995.

Schumer, P.D. "The Magician of Budapest." *Math Horizons* April, 1999: 5–9.

Schumer, P.D. "The Josephus Problem: Once More Around." *Math. Magazine* Vol. 75, No. 1, February, 2002: 12–17.

Index

Mathematical Journeys, by Peter D. Schumer
ISBN 0-471-22066-3